Optimizing Performance *of* Energy Systems

Optimizing Performance *of* Energy Systems

Saul Stricker, P. Eng.

BATTELLE PRESS
Columbus • Richland

Distributed by

MACMILLAN PUBLISHING COMPANY
A DIVISION OF MACMILLAN, INC.
NEW YORK

COLLIER MACMILLAN PUBLISHERS
LONDON

Distributed by

MACMILLAN PUBLISHING COMPANY
A DIVISION OF MACMILLAN, INC.
866 Third Avenue, New York, N.Y. 10022

COLLIER MACMILLAN CANADA, INC.

Printed in the United States of America.

Library of Congress Cataloging in Publication Data

Stricker, Saul, 1941–
Optimizing performance of energy systems.

1. Power (Mechanics) 2. Energy conservation.
3. Heating—Equipment and supplies. I. Title
TJ163.9.S76 1984 621.402 84-18470
ISBN 0-02-948940-7 (Macmillan)

ISBN 0-02-948940-7 Macmillan Publishing Company, New York

To Dori, Aimee, and Michelle
—for your encouragement
and understanding.

Contents

Tables and Figures

TABLES

FIGURES

Preface

The current trend of increasing energy prices is a continuing incentive for designers to improve the efficiency of energy-conversion processes and harness new sources of energy. Since many of the uses of energy involve heat production and heat flow, the need to measure these items accurately and inexpensively is likely to continue to increase. Improved measuring methods are being used to assess and further improve the efficiency of heating equipment. Space and water heating systems and wood-burning appliances are being designed in great numbers, but the performance of many of them are not known with certainty because of the involved and lengthy evaluation processes usually required. In addition, the performance of a complete system depends on the way in which it is used and on the installed capacity. The simple performance-evaluation techniques described herein will assist in the optimization of system designs and in the development of more appropriate applications practices.

The recent high level of activity in the solar heating field has resulted in a myriad of system designs and operating modes. In most cases, system optimization has been pursued by conducting several cycles of alteration and observation. This is a time-consuming, expensive, and unsatisfactory process; over the period of time involved, major changes in system performance can occur because of changes in controls, deterioration of collector surfaces, changes in demand for heat, and other aberrations. Testing under such circumstances is unreliable.

The proper evaluation of a solar system requires controlled data collection (over several months), reduction, and interpretation. Field monitoring over long periods can be expensive and can be the major part of the cost of developing an improved product or documenting a "demonstration" project. This book outlines several monitoring techniques that are suitable for detailed or overall performance evaluation in the laboratory and in the field. The emphasis is on simplicity, low cost, and reduction of the amount of data handling required to obtain bottom-line results of system performance. Also described are methods of simulating field performance in the laboratory to reduce the cost and duration of the evaluation period and to provide a direct comparison between solar and conventional heating systems.

Implementation of the techniques described range in duration from a few minutes for "instantaneous" performance monitoring to several months for seasonal evaluations; in detail, they range from manual readings to magnetic recordings of pre-processed data for automatic handling by computer.

The information will be of interest to those responsible for assessing the performance of operating solar systems and to electric utilities investigating the impact that this new type of load will have on their systems. Workers involved in the design and evaluation of solar and heat-pump space-heating and water-heating systems, fuel-burning appliances, fuel-burning heating systems that incorporate storage capability, and absorption cooling systems also will find the material helpful.

Integrated-circuit devices that can be used to assemble low-cost monitoring packages are described. The use of microprocessors described offers further short cuts and improvements in data gathering, data display, and evaluation of system performance.

Techniques addressed in this book have been used over several years by the research staff at Ontario Hydro. The staff developed the techniques in their search for a more efficient use of energy for water heating and space heating, in evaluating the effectiveness and impact of solar heating equipment on electric power systems, and in providing design assistance and performance feedback to manufacturers of solar equipment. A great deal of credit for the material goes to the technicians, engineers, and support staff who carried out the laboratory and field work. I am particularly indebted to Terry Strack, P.Eng., for gathering much of the material; to Andy Andruchiw, Ozzie Zomers, and Lydia Witczak for assisting in the preparation of the graphics; and to Mrs. Ina Ashmead and Mrs. Lou Ann Giesa for their typing and patient and cheerful assistance in general. I would like to acknowledge the cooperation of Ontario Hydro for permission to use technical material and drawings from internal publications.

Introduction

MEASUREMENT OF ENERGY IN THE CONVERSION PROCESS

Energy in the form of heat is a sublime form of energy. It disperses from its source to the environment. No amount of insulation can prevent this heat transfer, and measurement is a relatively simple process. In cases where heat transfer takes place by conduction and mass flow, measurement becomes more difficult, because additional inputs, such as mass flow and temperature difference, must be quantified. Heat energy is generally produced in a conversion process whereby some other form of energy is converted to heat—in a natural-gas-fired furnace, for example, the heating value of chemicals in the gas is converted to heat during the combustion process (see Figure 1). In another example, radiant solar energy incident on a solar collector is partly reflected

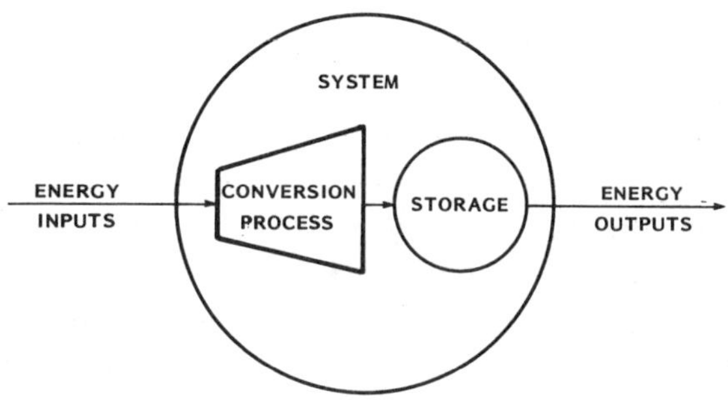

Figure 1. Energy flow into and out of a system.

by the glazing material, and the balance is converted to sensible heat on the absorber surface. A portion of the collected heat will be lost to the ambient air, and the balance is carried away by the heat-transfer fluid in the system.

In nature, there is always a perfect balance between inputs and outputs. In very general terms, the first law of thermodynamics asserts that the net flow of energy across the boundary of a system is equal to the change in energy of the system.[1] Whatever form the energy is in (solar, chemical, radiant, electrical, etc.), this law holds true:

$$\text{Energy Inputs} = \text{Energy Outputs}$$

unless heat is stored in the system, in which case:

$$\text{Energy Inputs} = \text{Energy Outputs} + \text{Heat Stored}$$

Consider the energy flows into and out of a house in a 24-hour period during the heating season. We will analyze the components of energy input and output, assigning numbers to them to enable us to visualize their relative magnitudes. Figure 2 presents cross-sections of the house. The top section indicates energy inputs to the structure; there are four:

1. Electricity
2. Fuel oil for space heating (being placed into a storage tank)
3. Heat energy given off by the occupants
4. Solar energy entering via the windows

The bottom section of Figure 2 indicates energy losses of the structure; there are seven:

1. Conduction of heat through walls (and foundation floors, neglected in this example)
2. Conduction of heat through ceilings
3. Conduction of heat through windows
4. Air leakage
5. Furnace exhaust lost to the atmosphere via the chimney
6. Hot water going down the drain
7. Appliance exhaust, such as that of a clothes dryer

There is always a balance between the energy inputs to the house and the heat lost by the house. The calculated magnitudes of energy inputs and heat losses are shown in Table 1 for a 24-hour period during the winter. (The assumptions for the case are also shown in the table.) Electricity consumption was obtained from electrical utility records[2] for February and adjusted for 92.9 sq. m (1,000 sq. ft.). Occupancy was three adults, full time, with 100 W heat output per per-

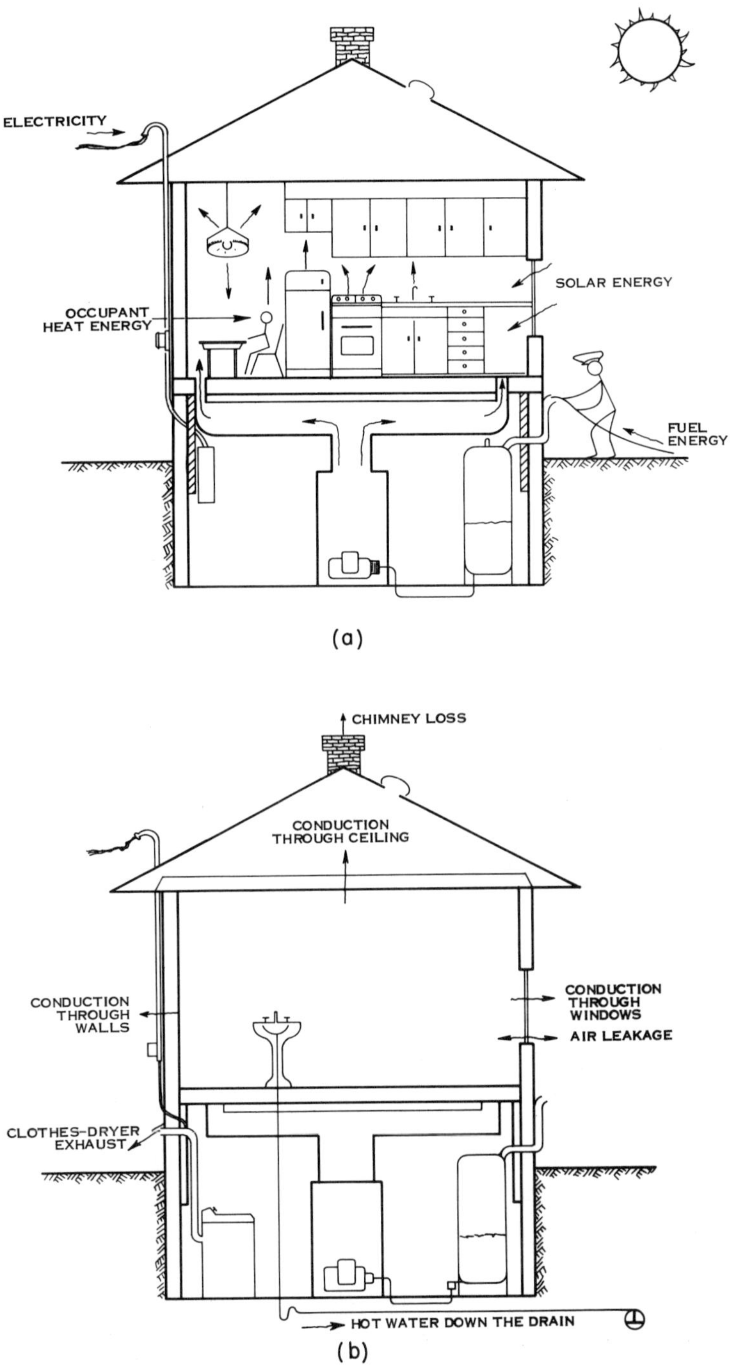

Figure 2. Energy inputs and heat losses of a house.

TABLE 1
Energy Input and Losses During a 24-Hour Period

Energy Inputs:		
1. Electricity (including water-heating)	104.4 MJ	29.0 kWh)
2. Fuel oil (7.5 U.S. gal. @ 60% efficiency)	658.8 MJ	(183.0 kWh)
3. Heat from 3 occupants	25.2 MJ	(7.0 kWh)
4. Solar energy input	25.2 MJ	(7.0 kWh)
	813.6 MJ	(226.0 kWh)

Energy Losses:	
1. Conduction through walls	151.2 MJ (42 kWh)
2. Conduction through ceilings	57.6 MJ (16 kWh)
3. Conduction through windows	90.0 MJ (25 kWh)
4. Air leakage	237.6 MJ (66 kWh)
5. Furnace exhaust to the atmosphere	230.4 MJ (64 kWh)
6. Hot water down the drain	32.4 MJ (9 kWh)
7. Clothes dryer exhaust	14.4 MJ (4 kWh)
	813.6 MJ (226 kWh)

Assumptions:		
House floor area	92.9 sq. m	(1,000 sq. ft.)
Exterior wall area	173.2 sq. m	(1,865 sq. ft.)
Total R-value	2.6 R(SI)	(R_t=14.7 hr.ft.^{2}F/BTU)
Ceiling area	92.9 sq. m	(1,000 sq. ft.)
Total R-value	3.6 R(SI)	(R_t=20.4 hr.ft.^{2}F/BTU)
Glass area (total)	13.9 sq. m	(150 sq. ft.)
South-facing glass area	5.6 sq. m	(60 sq. ft.)
R-value	11.1 W/m^2 °C	(1.95 BTU/hr. ft.^{2} °F)

All windows are double glazed.

Air changes:	
Kitchen and washroom	1½ air changes per hour (ach)
Living areas	¾ ach
Basement	½ ach
Design condition	−19.4°C (−3°F)

Average temperature during 24-hr. period of interest—4.4°C (24°F)

son.[3] The graphs of Figure 3 indicate the relative magnitude of the various components of inputs and losses. Note that there is a balance between the total inputs and the total losses.

In practice, quantifying the *inputs* is straightforward. Quantifying the *outputs* is usually more involved because of "losses" that may be difficult to isolate and measure. In the case of an oil furnace, for example, the inputs can be readily measured with conventional revenue metering of electricity and oil. The outputs are the heated air or water delivered to the premises, the heated flue gases expelled via the chimney, and the heat lost by the furnace and ductwork to the basement. Other minor components of heat inputs or outputs are also involved in the balance—the heat content of air provided for combustion, heat losses of the furnace flue piping to the house, and unburned fuel, for example. Measurement of heat energy and evaluation of the performance of solar and conventional heating and cooling appliances are the main subjects of this book. Although the applications described are in the domestic water-heating and space-heating areas, the methods may be applied to industrial processes involving heating and cooling as well.

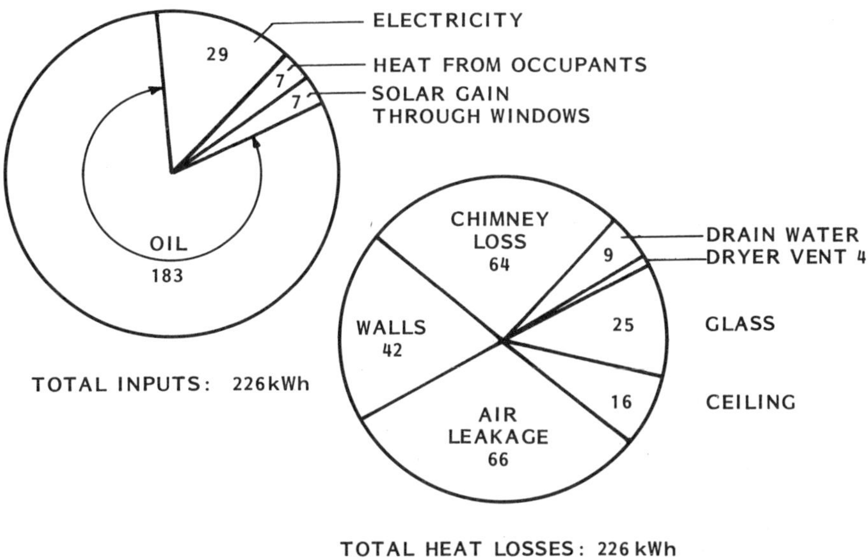

Figure 3. Components of energy inputs and heat losses of a house—24-hour period, winter conditions.

1

Measurement of Inputs

The measurement of energy inputs is generally straightforward, since it involves the use of common flowmeters developed for billing—kilowatt-hour (kWh) meters for electricity, gas meters for natural gas, weigh scales for coal or wood, and oil flowmeters for oil. Solarimeters can be used to measure the intensity of solar radiation. The amount of solar energy reaching the interior of a building is quantified from information about the size and angle of the windows, glazing material, and shading materials.[4]

The accurate determination of total energy inputs provides a figure against which the total of the energy outputs can be checked. Conversely, if accurate measurements of energy inputs and some outputs can be obtained, an indication of the remainder of the energy can be obtained by subtraction.

This chapter deals with the measurement of energy inputs. Subsequent chapters describe techniques for measuring and recording energy outputs in the form of heat in air-based and liquid-based systems.

MEASUREMENT OF RADIATION FROM SUN AND SKY

The radiant output of the sun is subject to small variations related to changes in the amount of ultraviolet radiation emitted. The periodicity of these variations is related to the frequency and size of the sunspots on the surface of the sun, which come and go in cycles related to 273 months. The maximum ultraviolet variation observed by Dr. Abbott[5] was ±1.8 percent. The highest values were observed when sunspot activity was intense in areas away from the solar equator; the lowest values were observed when large sunspots occurred at the equator.

The intensity of solar radiation incident on the earth is also a function of the distance between the earth and the sun. The earth moves in an elliptical orbit about the sun, which is stationed at one of the foci of the ellipse. The distance varies from the mean of 1.496×10^8 km (one Astronomical Unit) by ±1.6 percent. The solar radiation intensity incident on the earth varies by ±3.4 percent.

The intensity of radiation at high altitudes has been measured by means of balloons, rockets, and the Mariner-Mars probe. The most probable value of the mean solar radiation intensity outside the earth's atmosphere is 1353 Wm^{-2}, called the Solar Constant I. The spectral distribution of this radiation is similar to that of a blackbody at 5762 K and ranges in wavelength from about 0.2 μm to more than 2.6 μm, with a maximum radiation of between 0.4 μm and 0.5 μm. As this radiation passes through the atmosphere, some energy is absorbed by ozone, oxygen, water vapor, and carbon dioxide. Consequently, the intensity and spectral distribution of the radiation that reaches the surface of the earth has been affected by a number of factors in addition to local cloud and air particulate levels. At sea level, with a very clear atmosphere, the solar and sky radiation density is about 1110 Wm^{-2}, or 18 percent lower than the level outside the atmosphere.

The amount of radiation received on a surface at ground level will be affected not only by the transmission characteristics of the atmosphere, cloud cover, and suspended particulates in the air but also by the angle of the surface relative to the sun's position.

The radiation reaching the surface has two components: one received directly from the sun, the other indirectly via diffusion from clouds, sky, and particulate matter. Solar collectors that use concentrating devices such as mirrors and lenses are capable of concentrating the direct component only. On cloudy days, when the direct component of solar radiation is very low, these concentrators are ineffective. Collectors that do not rely on concentrators, such as the flat-plate type, are capable of absorbing both direct and indirect radiation components. Depending on the type of solar collector being studied, it may be necessary to record both the total and indirect components of solar radiation.

TOTAL AND INDIRECT RADIATION

Normally, devices used for measuring solar radiation are of the type that respond to total—direct and indirect—sky radiation. These devices must be located at sites that will not provide shading or reflections from roofs, walls, windows, etc. during the entire year. The indirect component of radiation may be measured by a second device equipped with an adjustable shadow ring set to cast a shadow over the sensing area at all times. In this manner, the sensor is exposed to indirect radiation from the whole sky except that insignificant portion blocked by the shadow ring. The direct component can be determined readily by subtracting the indirect radiation from the total.

MEASURING TOTAL RADIATION

Total radiation is measured by means of a calibrated thermopile consisting of blackened hot-junction receivers and whitened cold-junction receivers enclosed in a glass hemisphere transparent in the range of 0.28 μm to 2.8 μm. These devices are called pyranometers or solarimeters. Two commercially available units are illustrated in Figure 4.

Solarimeters produce an electrical output proportional to the incident radiation—typically in the 0.15 mV range. This output can be recorded by circular or strip-chart recorders and integrated over time (hourly or daily, for example) by graphic or mechanical methods. When a graphic record is not required, integration may be accomplished by means of an analog to pulse-rate converter and a counter as illustrated in Figure 5. In such devices, the linear millivolt signal generated by the solarimeter is amplified and then converted to a series of pulses whose frequency is directly proportional to the amplitude of the voltage. Typically, the converter has a maximum output pulse rate of 10 kHz, which is divided by a "divide-by" circuit to accommodate the maximum input pulse rate of the counting device—1 pulse per second for an electromechanical counter, for example. Printing counters programmed to print each quarter hour, hourly, or daily can be used for automatic integration and recording.

RECORDING SOLARIMETER OUTPUT

A variety of methods for recording, processing, and presenting solari-

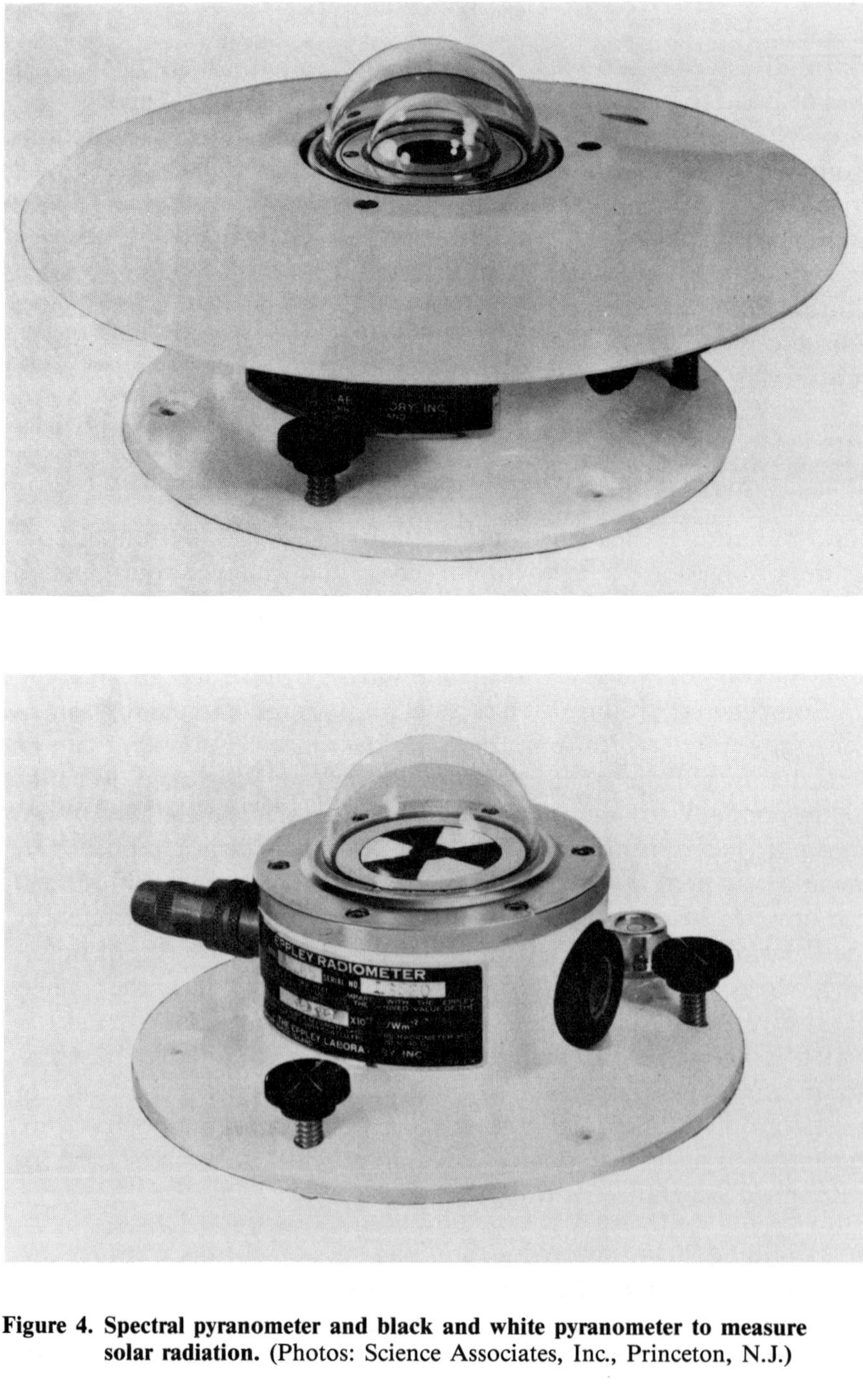

Figure 4. Spectral pyranometer and black and white pyranometer to measure solar radiation. (Photos: Science Associates, Inc., Princeton, N.J.)

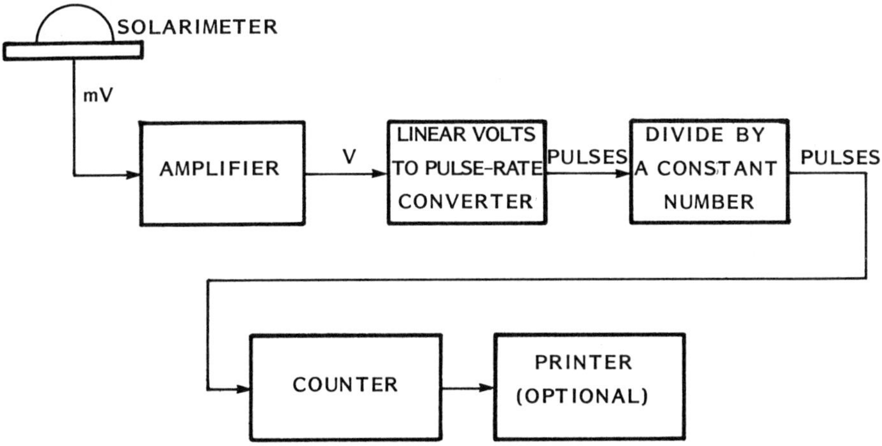

Figure 5. Arrangement to integrate solarimeter output.

meter outputs are available, and considerable effort can be saved by selecting the method most appropriate for the intended use. Generally, the cost of processing and handling the data will be lowest when the maximum amount of processing is carried out prior to recording. For example, if total daily insolation is to be measured, it would make more sense to integrate the output of the sensor electronically and take a single daily reading (by hand or automatic printer) than either to record the output continuously on a chart and integrate by hand or to record periodic instantaneous readings digitally and process the records later. On the other hand, when virtually instantaneous output records are required, integration over short time periods, such as 5 minutes or 15 minutes, should be undertaken and the results recorded. Because of the highly variable levels of sunshine, particularly during partly cloudy weather, instantaneous recording by digital methods requires a sampling rate of several dozen readings per minute if serious error is to be avoided. This type of data recording generates a vast amount of information and requires a considerable amount of storage capability and computer time. Data loggers with magnetic storage and millivolt input sensitivities are suitable for this type of recording.

"Digital demand recorders" or "load survey recorders," which are frequently used by electric utilities for power-use surveys and billing, are particularly useful devices for recording solarimeter information in pulse form on a time base. These recorders use a slow-moving ¼-inch magnetic-tape cartridge on which data pulses are recorded on one or more tracks while timing pulses (every 15 minutes, hourly, etc.) are recorded on another track. Normally a single cartridge can record over one month's data. A photograph of a commercially available 4-track recorder (3 data tracks and 1 timing track) is shown in Figure 6. A battery carryover to ensure continuity of the time base during power outages is available. The minimum time between data pulses depends on the tape speed selected; for a 36-day record, a minimum time of 0.6 seconds is required between data pulses; for a 108-day record, 1.8 seconds are required. To retrieve information from the tape, the number of data pulses within each time sequence is counted electronically and the results, identified by time segment, written onto computer tape, data cards, or printer paper. Such a retrieval facility is shown in Figure 7.

Another type of demand recorder uses a microprocessor and a solid-state memory cartridge with capacity sufficient to store counts at 15-minute intervals from four data channels for a period of one month. A photograph of this type of recording device is shown in Figure 8. The associated microprocessor-based reader/computer (translator) is shown in Figure 9. This device reads the data contained in the cartridge and prepares it for print-out in the most useful format or for interface to a main-frame computer. A second example of a microprocessor-based recorder and a compatible translator are presented in Figures 10 and 11, respectively.

FUEL INPUTS

MEASUREMENT OF ELECTRICAL ENERGY

Quantification of the electrical energy (E) used by a device involves computing the product of current (I), voltage (V), and operating time (t):

$$\text{Electrical energy used} = \text{Voltage} \times \text{Current} \times \text{Operating time}$$

This basic equation is valid only for steady values of voltage and current and for purely resistive loads. It is valid for both alternating current (ac) and direct current (dc); for ac current the voltage and cur-

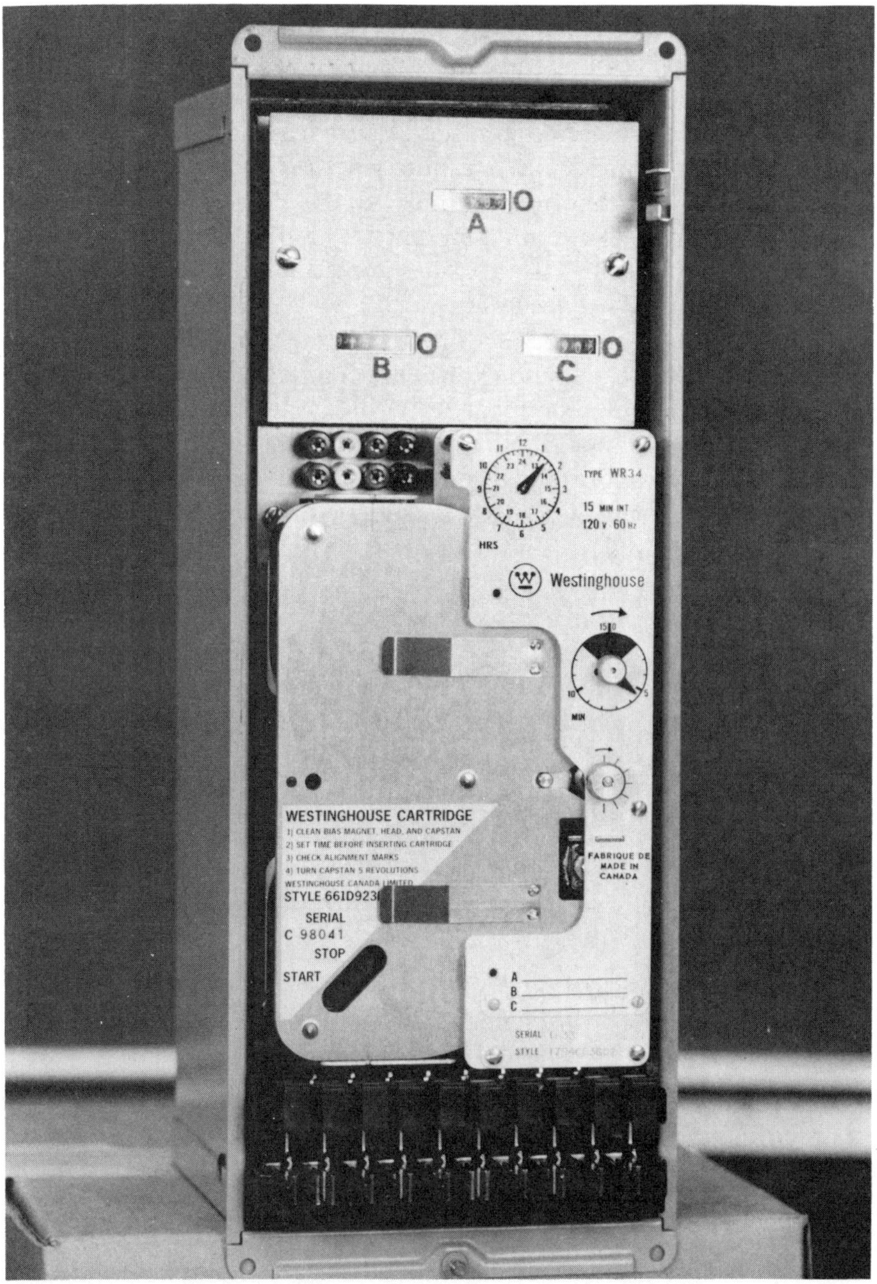

Figure 6. Magnetic tape recorder; 3 data tracks and 1 time track.
(Photo: Westinghouse Canada Ltd.)

Figure 7. Translating facility for magnatic tape records of pulse data. (Photo: Westinghouse Canada Ltd.)

Figure 8. Microprocessor-based recorder. (Photo: EMAX Inc. and OLMAN Engineering, Canada.)

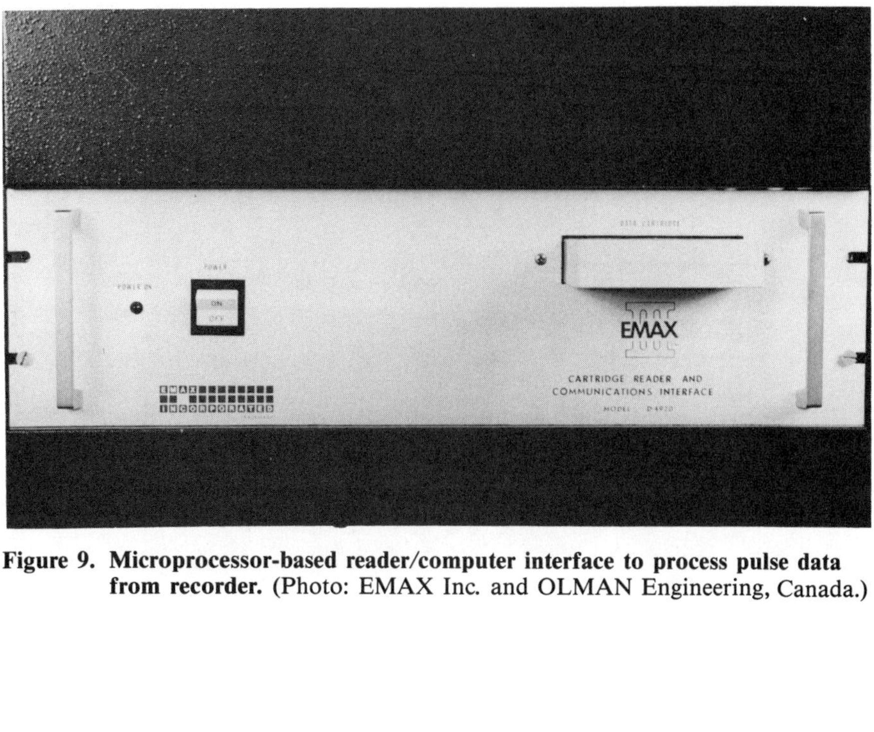

Figure 9. Microprocessor-based reader/computer interface to process pulse data from recorder. (Photo: EMAX Inc. and OLMAN Engineering, Canada.)

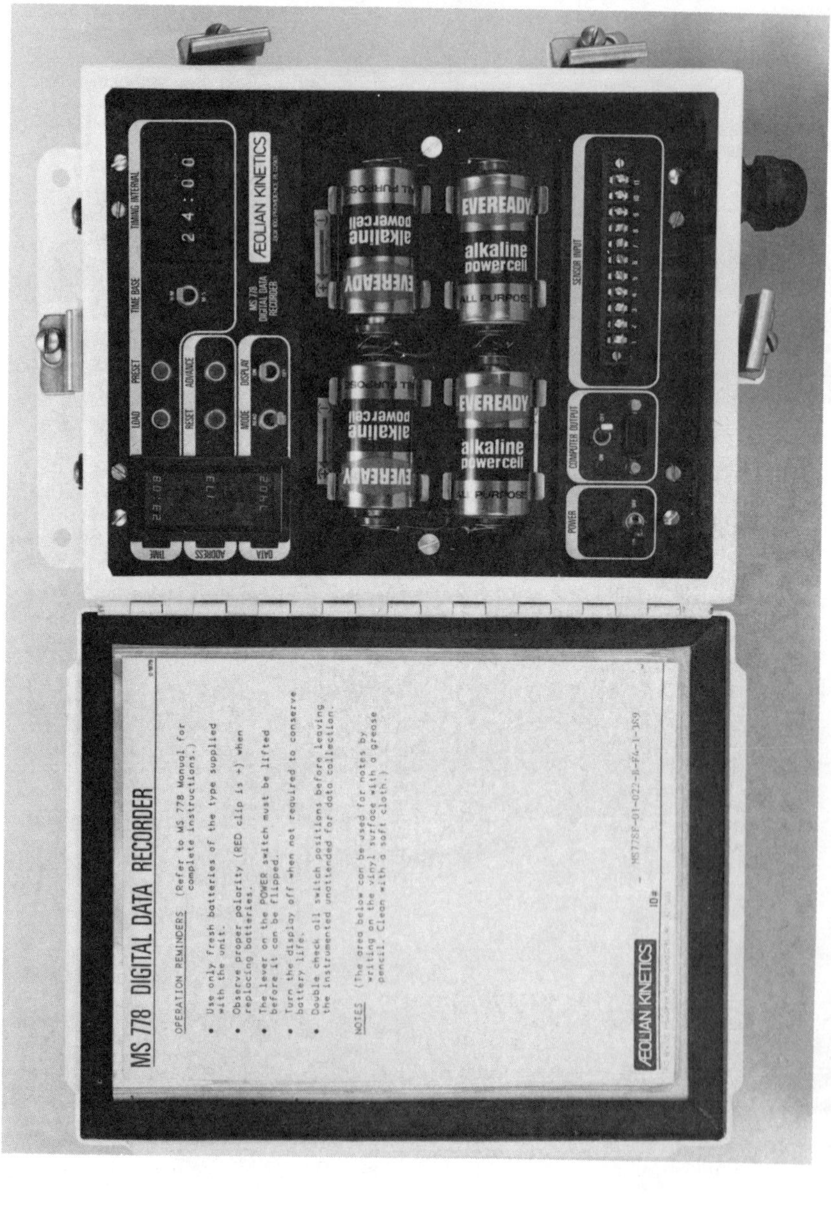

Figure 10. Microprocessor-based recorder. (Photo: AEOLIAN KINETICS, Providence, R.I.)

Figure 11. Microprocessor-based programmable data logger/translator. (Photo: AEOLIAN KINETICS, Providence, R.I.)

rent must be expressed as the root-mean-square values. In terms of units, the equation reads:

$$\text{Kilowatt hours} = \frac{\text{Volts} \times \text{amperes} \times \text{hours}}{1000}$$

If the current or voltage varies with time, then the electrical energy used is the integral of the instantaneous product:

$$E = \int_0^t V(t)\ I(t)\ dt$$

In cases where the electrical load is inductive or capacitive (as is the case with motors and ballasted gas discharge lamps), the cosine of the phase angle between the voltage and current, θ, must be introduced into the equation, as follows:

$$E = \int_0^t V(t)\ I(t)\ \cos\theta\ dt$$

Commercial billing kWh meters perform this computation electromechanically—very accurately, but *only* for alternating current—and store the total energy value in a set of dials. Such meters are available for a variety of voltages, in single, dual, and three-phase versions, and for use in conjunction with current transformers to handle loads over 200 amperes. Some are available with outputs to operate counters, printers, or demand recorders of various types.

Recording energy flow in systems that use direct current requires the use of a wattmeter, which measures the product of voltage and current but does not integrate with time. The integration must be done by hand from graphic records or by converting the analog output to a pulse train and using a counting/totalizing device as described previously.

MEASUREMENT OF GAS ENERGY

Quantification of gaseous-fuel input requires that the mass (quantity) of fuel consumed and its calorific value, (heat of combustion) be measured.

Measurement of the quantity of fuel used can be performed by commercial billing meters. Some models are available with pulse-output capability; others, with remote-reading capability, may lend themselves to automatic periodic recording. Where bottled gas is used, periodic weighing of the bottle may suffice. If the pressure at the burner is stable, it may be practical to calibrate the nozzle flow rate and assume that the flow remains constant. In this case, a record of the control valve's ON time may be sufficient to establish the amount of fuel used. A convenient method of recording ON time is to have a motor-driven switch produce pulses while the burner is operational. The train of pulses can then be totalized or recorded in a manner similar to that described previously for recording solarimeter output. The frequency of the pulses can be selected to be compatible with the recording medium selected. For example, if the motor-driven switch operates at a rate of one ON-OFF operation per second, a 120-count record would indicate a 2-minute operation of the burner. By calibrating the burner nozzle in terms of weight or volume of fuel burned per hour, each recorded count can be converted to a specific amount of fuel used.

The calorific value of the fuel may be determined by measuring the heat evolved during combustion of a known quantity of the fuel in a calorimeter, or it may be estimated by chemical analysis of the fuel and application of the known heating values of the several chemical elements present. Fuels that contain hydrogen include water vapor in the products of combustion. If the vapor is allowed to condense during the calorimeter test, the resulting heating value is called the "higher [or gross] heating value." If the vapor is not allowed to condense, the resulting heating value is called the "lower [or net] heating value." For combustion appliances, the gross heating value of the fuel consumed is normally quoted. Table 2 lists volumetric calorific values of common gaseous fuels. Net heating values are about 9 percent lower than gross heating values for natural gas, and 8 percent lower for propane and butane.

MEASUREMENT OF LIQUID FUEL ENERGY

Liquid fuels derived by refining crude petroleum are mixtures of hydrocarbons. Specifications established for their properties subdivide them into various grades, as shown in Table 3. Liquid-fuel consumption may be measured by a commercial oil-billing meter and recorded by periodic manual readings of the meter. Some meters can be equipped with a pulse initiator, which provides a convenient output for driving

TABLE 2
Gross Heating Values of Gaseous Commercial Fuels[6]

Natural Gas	MJ/m³	BTU/ft³ (dry)
High inert	35.7 to 39.2	958 to 1051
High methane	37.6 to 39.9	1008 to 1071
High BTU	39.9 to 41.8	1071 to 1124
Propane (5 to 10% propylene)	93.2	2500
Butane (5% butylene)	119.2	3200

Reprinted with permission from the American Society of Heating, Refrigerating and Air-Conditioning Engineers, Inc., Atlanta, GA.

a printing counter, a pulse totalizer, or pulse-recording equipment for automatic data handling and processing. Fuel consumption can also be monitored by recording ON time of the burner and calibrating the flow through the nozzle at the operating pressure of the burner. Measurement of oil flow into space heaters fed by gravity may not be possible with commercial oil-flow meters. In such cases, a convenient means of measuring the rate of consumption is to note the exact mass of oil required to fill the tank and then observe the length of time the burner operates between fill-ups.

MEASUREMENT OF SOLID FUEL ENERGY

Combustion equipment fired with solid fuel does not generally lend itself to convenient methods of measuring the rate of fuel consumption. One of the most practical methods is to weigh the fuel prior to use and to keep a regular log of each charge. The heating values of common solid fuels are given in Table 4.

TABLE 3
Gross Heating Values of Liquid-Petroleum-Derived Fuels[7]

Grade of Fuel Oil	Properties & Application	Density		Gross Heating Value	
		kg/1	lb/ US gal	MJ/1	BTU/ US gal
No. 1	Suitable for vaporizing a pot-type burner or spray burner.	0.8329 to 0.8000	6.950 to 6.675	38.18 to 37.04	137,000 to 132,900
No. 2	Oil for general purpose heating.	0.8744 to 0.8341	7.296 to 6.960	39.52 to 38.18	141,800 to 137,000
No. 4	Preheating is not usually required for handling and burning.	0.9332 to 0.8863	7.787 to 7.396	41.28 to 39.88	148,100 to 143,100
No. 5 (Light)	Preheating may be required for handling and burning depending on weather and application.	0.9515 to 0.9211	7.940 to 7.686	41.81 to 40.92	150,000 to 146,800
No. 5 (Heavy)	Preheating may be required for burning and in cold climates may be required for handling.	0.9683 to 0.9455	8.080 to 7.890	42.36 to 41.64	152,000 to 149,400
No. 6	Preheating is required for handling and burning.	1.0124 to 0.9651	8.448 to 8.053	43.45 to 42.17	155,900 to 151,300

Reprinted with permission from the American Society of Heating, Refrigerating and Air-Conditioning Engineers, Inc., Atlanta, GA.

TABLE 4
Gross Heating Value of Solid Fuels[8]

Fuel	Heating Value	
	MJ/kg	**BTU/lb**
Anthracite coal	29.54	12,700
Bituminous coal	25.60–33.40	11,000–14,350
Lignite	16.00	6,900
Various woods (with 20% moisture)	16.00	~6,900

Reprinted with permission from the American Society of Heating, Refrigerating and Air-Conditioning Engineers, Inc., Atlanta, GA.

2

Measurement of Heat Output—Principles

DIRECT METHOD

Heating and cooling appliances, such as furnances, air conditioners, and water heaters, are controlled by cycling their input energy sources ON and OFF. As a consequence, the energy output is intermittent and variable. In the case of forced-air furnaces and air conditioners, the air flow is intermittent but relatively constant when the appliances are ON; the temperature is variable and cyclic. Measuring the energy output of heating appliances presents a number of problems because of the variable nature of the quantities to be measured.

The rate at which thermal energy is picked up by a stream of air or water flowing through a heating unit (see Figure 12) can be quantified by computing the product of the rate of mass flow (throughput), the average bulk temperature rise between outlet and inlet, and the specific heat of the fluid. In consistent units the equation is as follows:

$$\frac{\text{Heat output}}{\text{unit time}} = \frac{\text{mass flow}}{\text{unit time}} \times \text{temperature-rise} \times \text{specific heat} \tag{1}$$

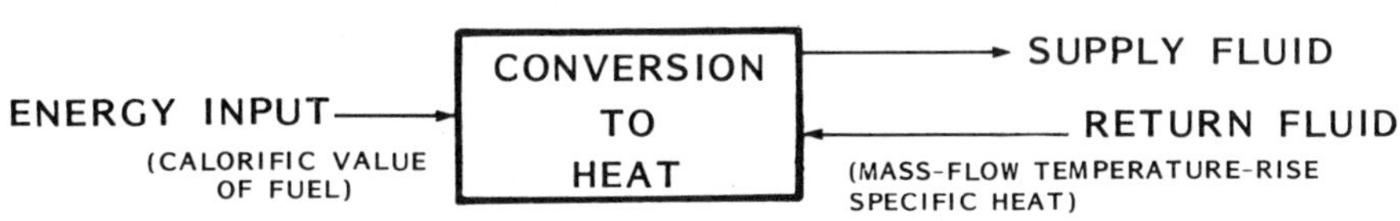

Figure 12. Conversion of fuel energy to heat.

For forced-air furnaces, it may be difficult to obtain an accurate measure of the rate of air flow. In addition, because of the shape of the heat exchangers, the temperature of the airstream leaving the furnace may be highly non-uniform.

In hydronic heating systems or water heaters, the flow of water and the temperature distribution across the face of the stream in pipes is generally better behaved than in forced-air furnaces. As a consequence, the determination of rate of heat flow is usually more straightforward.

AIR SYSTEMS

The flow of air in ducts may be turbulent or streamlined with little lateral mixing, depending on the duct configuration, distance from elbows, surface finish, and air velocity. In long straight runs of more than 8 duct diameters from an elbow, the air flow in well designed ducts used for heating and cooling of buildings is streamlined, well behaved, and subject to little lateral mixing. The velocity distribution across the face of the airstream is nearly parabolic, as indicated in Figure 13 for a round duct.

The average velocity of an airstream in a duct can be calculated by subdividing the cross-section into several equal areas, determining the velocity at the center of each area, and computing the average of all the velocity values. Air velocity at a specific point in the air stream can be determined by stationing a pitot tube at the point of interest and connecting it to a manometer.[9] This arrangement provides a reading of the differential pressure between the velocity pressure at the specific point and the static pressure. The relationship between velocity and pressure is as follows:

$$V = C\left(\frac{h_w}{\rho}\right)^{1/2} \qquad (2)$$

where V = velocity in m/s (ft./min.)

h_w = velocity pressure (pitot tube manometer reading) in mm (inches) of water

ρ = density of air, in kg/m³ (lbs./cu. ft.)

C = 4.42 (1096.5)

The acceptable velocity range for using this method with a micromanometer (at accuracies of 1–5 percent) is 1–50 meters per second (180–10,000 feet per minute). In a duct of rectangular cross section,

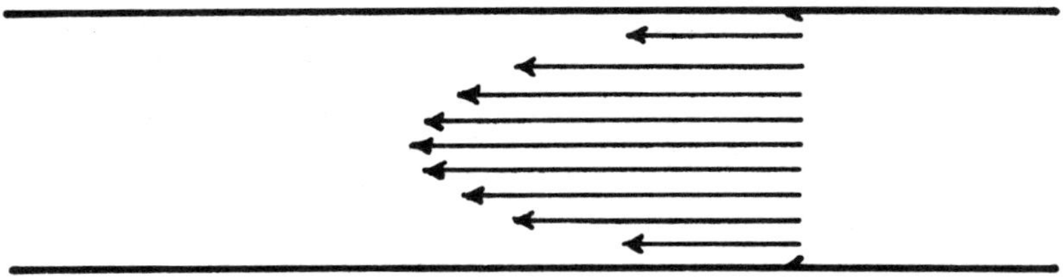

Figure 13. Velocity distribution of moving air in a duct.

the integration can be carried out by subdividing the cross section into 20 or more equal-area rectangles; a circular duct can be subdivided into annular sections of equal area (see Figure 14). In circular ducts, readings should be taken along two diameters, as indicated in the figure.

The measurement of the bulk temperature of air moving into a duct is straightforward when the air temperature is uniform across the entire cross section. This is often the case in the return-air duct. In the case of a supply-air duct, wherein the air temperature is not uniform across the whole cross section, the measurement is more complicated. This situation requires that velocity and temperature be measured simultaneously at sufficient points to permit calculation of a weighted average temperature. The usual procedure is to select a section of duct where the flow is laminar and well behaved, divide the duct cross section into areas as shown in Figure 14, and measure velocity (V) and temperature (T) at the center of each area.

The weighted average temperature is the sum of the products of V and T at each area divided by the sum of the Vs for all areas. In general terms:

$$\text{Average bulk temperature} = \sum_{\substack{i=1\\j=1}}^{\substack{i=5\\j=5}} V_{ij}T_{ij} \div \sum_{\substack{i=1\\j=1}}^{\substack{i=5\\j=5}} V_{ij} \qquad (3)$$

This procedure is practical when the air temperatures do not vary with time; the method is so time-consuming that any variations in temperatures during the process would introduce serious errors. Since in most applications the temperatures *do* vary with time, some method of measuring bulk temperature is required. The most logical procedure would be to eliminate the need for making the multiple readings in

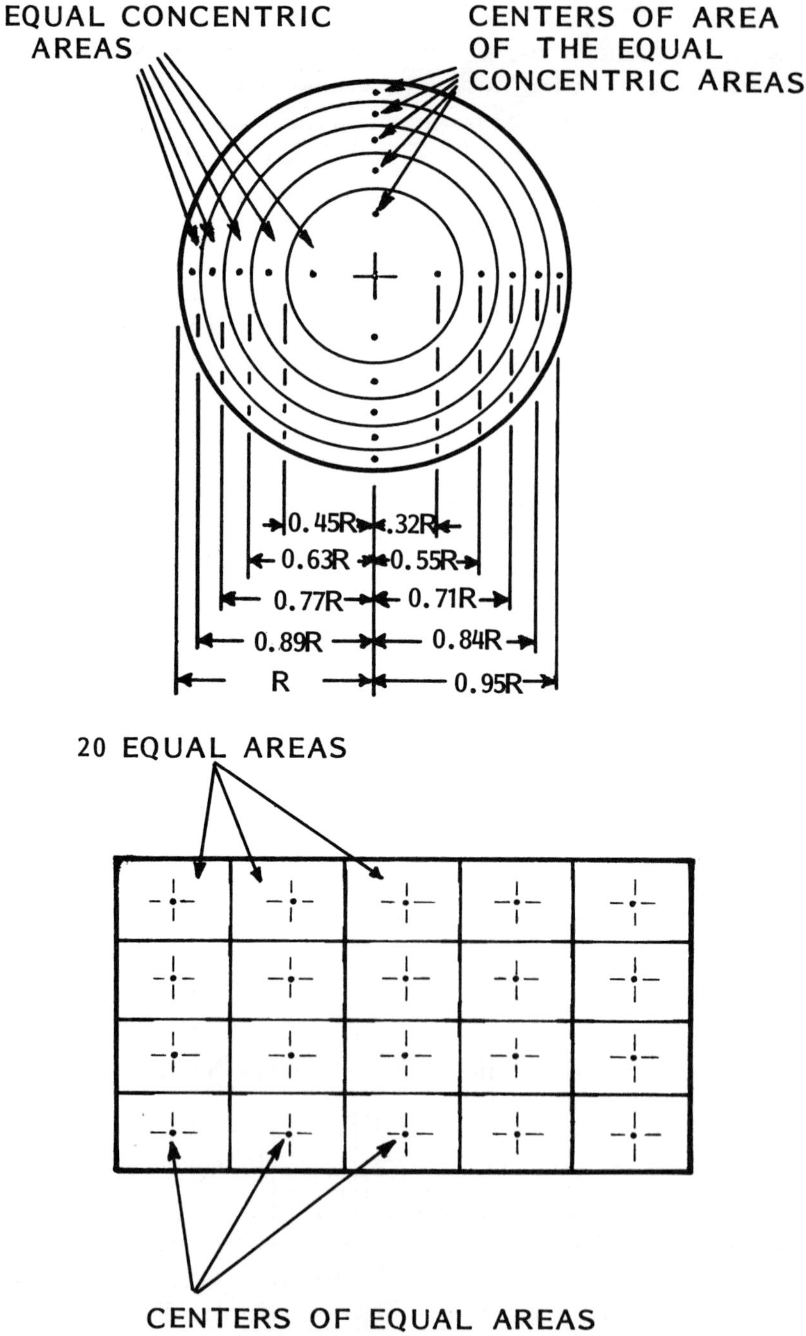

Figure 14. Measurement of bulk temperature in air ducts where air velocity and temperature are variable.

the first place—i.e., cause the airstream to have a uniform temperature across the whole duct cross section. Adequate mixing can produce this result and thereby permit the use of a single temperature probe at a fixed location. Mixing must be done in a turbulent section of duct or by means of a motor-driven mixer, because little lateral mixing takes place once laminar flow becomes established. One further caution: heat loss (or gain) from a duct wall affects the temperature of air near the duct wall. This aberration can be reduced by insulating the section of ductwork involved in the measurement process. This most desirable arrangement for measuring bulk temperature rise is shown in Figure 15.

The quantification of energy flow is now reduced to measuring the temperatures at the return and supply points (locations A and B in Figure 15) and the rate of mass flow and exercising equation 1:

Energy flow = (mass flow) × (temperature at B − *temperature at* A) × (specific heat)

where mass flow = (average air velocity) × (air density) × (duct cross-sectional area)

Mass flow can be quantified by subdividing the cross section of the duct into equal areas and measuring air velocities as explained previously or, more simply, by installing orifice plates or nozzles and measuring pressure drops that correspond to known air flows.[10] The simpler method can be used because the outlet airstream now is at a uniform temperature equal to the velocity-weighted average temperature.

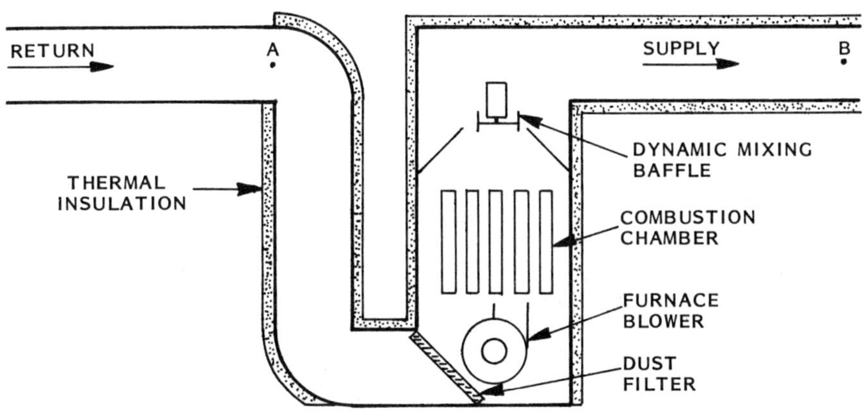

Figure 15. Arrangement for measuring bulk temperature rise across a heating duct.

LIQUID SYSTEMS

Hydronic space-heating systems with forced circulation can be instrumented readily to measure the rate of heat output. The two basic values, mass flow and bulk temperature rise, can be obtained with a commercial water-flow meter and two immersion-type temperature sensors, one installed in the return line and one installed in the supply line.

Figure 16 presents a typical hydronic furnace and circulating pump equipped with a water meter and temperature sensors at A and B. Insulation is shown installed on the piping between A and B to reduce heat gain or loss in the piping. Measurement of the rate of heat output consists of recording the temperature difference, ΔT, between points A and B over a fixed period of time, t_O, and noting the volume of water that circulated, V_O, over the time period:

$$\text{Energy delivered} = V_O \times (\text{average } \Delta T \text{ over time period } t_O) \times (\text{specific heat of water}) \quad (4)$$

$$\text{Average power output} = \frac{\text{Energy delivered}}{t_O}$$

Space-heating furnaces operate for periods of several minutes about one–six times per hour; the circulating pumps operate continuously in some systems and cycle on a thermostat in others. In either case, the water flows at a constant rate, Q_O, or not at all. This feature permits the use of the average temperature rise of the furnace, ΔT, over

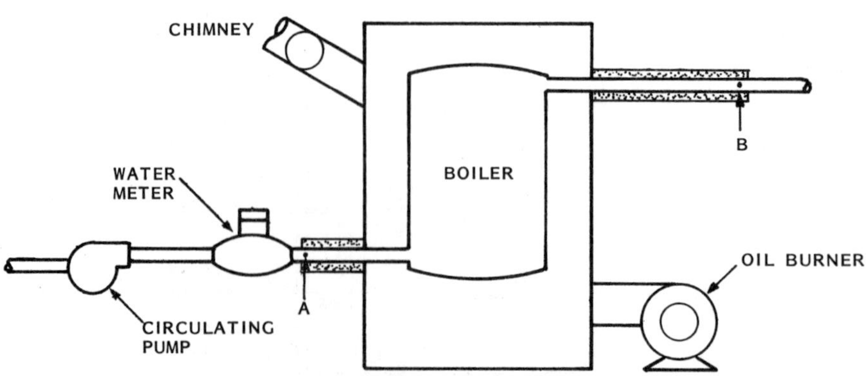

Figure 16. Hydronic furnace with temperature sensor locations shown at A and B and water flowmeter installed.

the period of operation of the pump, t_o, in the calculation of rate at which energy is being delivered. Figure 17 shows the constant rate of flow versus time and the temperature rise across the furnace during a normal cycle, as represented by the following equation:

$$\text{Energy delivered} = (\text{Specific heat}) \times Q_o \int_0^{t_o} \Delta T_o \, dt \tag{5}$$

$$\text{Average power output} = \text{Energy delivered}/t_o$$

This quantification may be conducted over any period of time—1 minute, 15 minutes, 24 hours, or an entire heating season.

The fact that water is circulated at a constant rate makes the quantification of energy flow in space-heating furnaces relatively straightforward. By constrast, quantification of the system energy output of

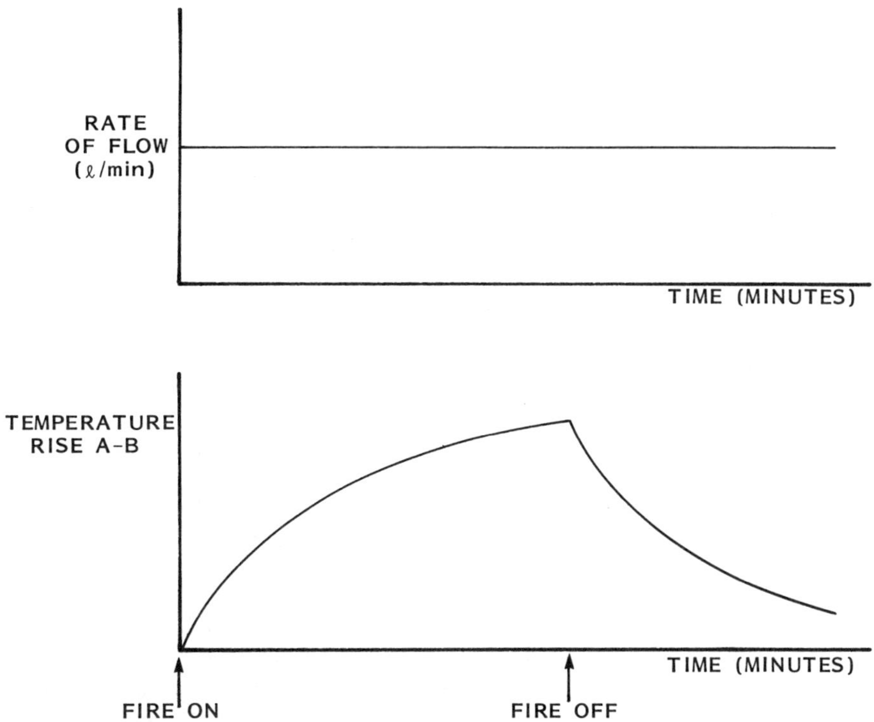

Figure 17. Water flow rate and temperature rise versus time for a typical hydronic furnace.

domestic-service water-heating systems is fraught with problems because of the highly non-uniform flow of water. These systems operate in a totally different fashion than space-heating systems—water flows usually are intermittent, irregular, and of varying rates due to the nature of demand for hot water. The temperature rise between inlet and outlet of a domestic-service water heater can vary considerably with time, depending on the fuel source. Gas- and oil-fired water heaters tend to have high firing rates, flues that cause temperature stratification vertically in the tank, and controls with high operating differentials. As a consequence, outlet water temperature may vary considerably, depending on the time of the draw relative to an ON period of the flame. Electric water heaters with dual elements generally exhibit less stratification and more uniform outlet temperatures because of the smaller differential between ON and OFF setting of the thermostat and the water mixing that takes place in the tank as a result of the operation of the element located near the bottom.

The upper graph of Figure 18 illustrates hot water draws on a time base in liters per draw. The lower graph illustrates the behaviour of the temperature difference ΔT across the water heater. It is virtually constant during a draw and may drop during long draws, depending on the amount of hot water in storage and on the temperature gradient in the tank. Between draws, ΔT drops because of the tendency of the cold water intake pipe to warm up and the outlet pipe to cool down. The effect can be reduced somewhat by locating the outlet temperature sensor as close as possible to the tank and by insulating the cold water inlet pipe. Another phenomenon tends to raise the temperature of the cold water inlet: as the water heater recovers, water expands and backs up into the inlet pipe, thus raising the temperature of this pipe and lowering the temperature rise across the tank. Another method of ensuring that the cold water inlet does not heat up is to install a bleed for the cold water line between the temperature sensor and the inlet to the tank. The water flow meter must be installed downstream of the bleed in order to prevent its registering the flow of water being bypassed.

To quantify the energy added by the water heater to the incoming water, the temperature rise between outlet and inlet and the volume of each draw must be measured. Energy is quantified as previously described by multiplying the volume of the draw by the temperature rise across the system and by the specific heat of water. One set of instrumentation that can be used is a water meter with pulse output

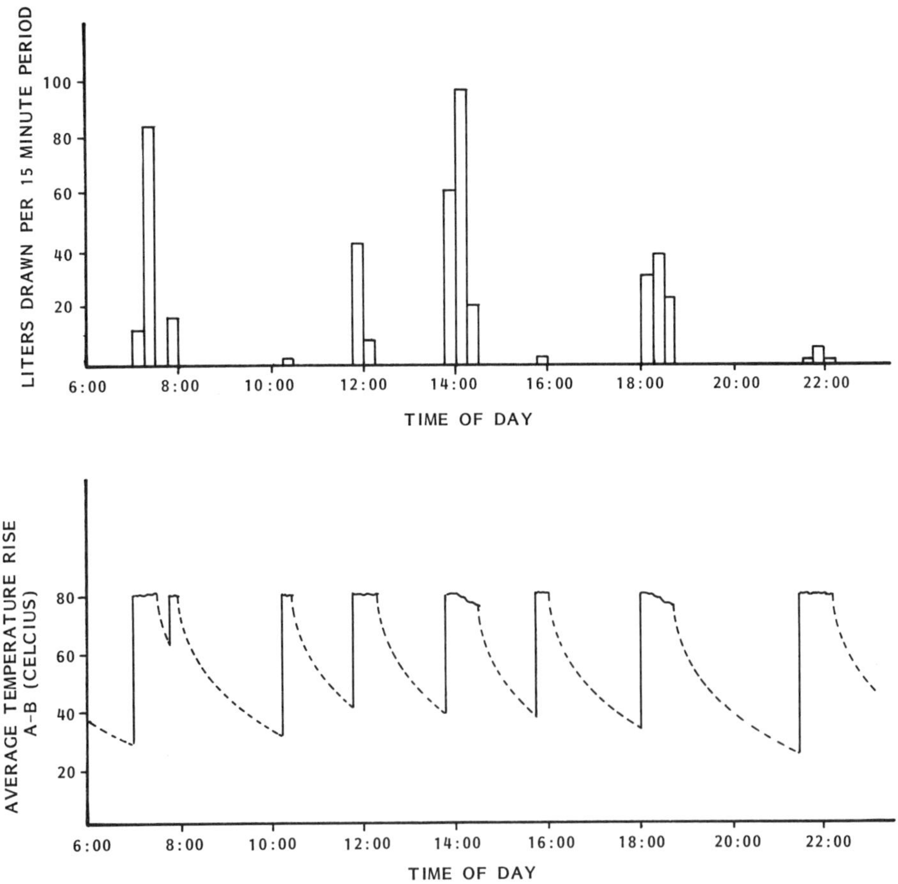

Figure 18. Water flow rate and temperature rise versus time for a typical domestic water heater.

and two temperature sensors located at A and B as illustrated in Figure 19. From Equation 4:

$$\text{Energy} = Q \times (\text{duration of draw}) \times \Delta T \times (\text{specific heat of water})$$

where: Q is flow rate in volume per unit time
ΔT is temperature rise across A-B

In most cases, flow rate and temperature rise vary simultaneously, so the product of volume and temperature rise must be obtained a number of times sufficient to provide an acceptable resolution of the

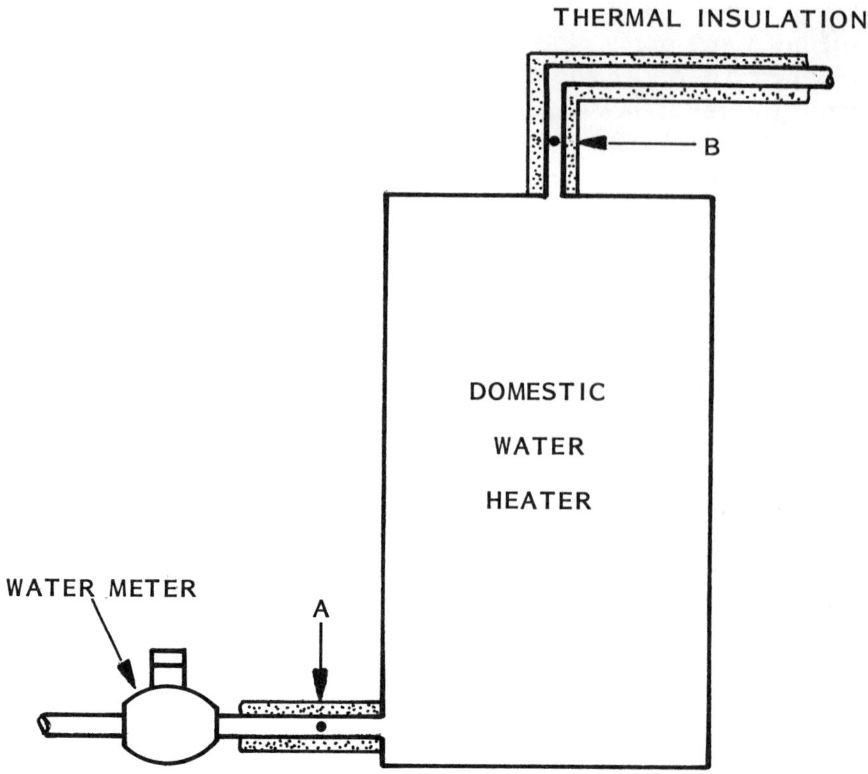

Figure 19. Domestic water heater with water-flow meter installed; locations of temperature sensors shown.

quantities of interest. For example, if a resolution of one liter of water is selected, then the average temperature of each liter of water drawn must be obtained.

Ideally, the mean temperature of each incremental volume is required, but in practice this may not be feasible; a spot measurement taken sometime during the delivery of a given volume may have to suffice. Knowing the rate at which water temperature may change in the installation and the selected increment of volume, the maximum expected error can be calculated. In water heater installations, where the draws are not sufficiently long to deplete the storage, the changes in temperature rise between outlet and inlet are relatively small. However, at the beginning of each draw, depending on the length of time since the previous draw, the temperature difference will be considerably lower because of warm-up of the cold water supply pipe and cool-down of the hot water outlet pipe. For the first few seconds, the temperature rise between outlet and inlet will be low until cold water at mains temperature reaches the tank. Temperature sensors that have a fast response (on the order of 15 seconds or less) should be selected to reduce the introduction of errors during periods of quickly changing temperature.

For example, the sum of the temperature-rise values observed individually for each liter of water drawn is equal to the *volume-weighted temperature rise* multiplied by the number of liters drawn. By multiplying this number by the specific heat of water, one obtains the total energy added to the water by the water heater. Very convenient methods of recording temperature rise (and energy) can be used if a signal voltage whose amplitude varies directly with temperature difference is converted to a pulse rate as described in the section on recording solarimeter output in Chapter 1. The pulses then may be counted for a fixed period (such as one second) each time a unit volume (such as one liter) is drawn from the water heater. For example, if a temperature difference of 50 degrees causes 50 pulses to be produced in one second and zero temperature difference causes zero pulses to be produced, then, each time a liter of water is drawn, the number of pulses recorded in one second will equal the temperature rise in degrees (assuming the relationship is linear). By multiplying the pulse count by the specific heat of water, the energy added by the water heater is quantified. The energy equivalent of one pulse for this example is as follows:

$$\frac{\text{Energy content of 1 liter of water raised } 50°\text{C}}{50 \text{ pulses}}$$

$$= \text{Volume} \times \text{temperature rise} \times \text{specific heat} \div 50$$
$$= 1 \times 50 \times 0.004187 \div 50 = 0.04187 \text{ MJ/pulse}$$

Gating temperature pulses with increments of volume provides, in effect, the product of temperature rise and flow—a technique called "pulse-gating multiplication." Not only is the hardware required to multiply the two variables simple and inexpensive, but also the energy flow information, being in pulse form, can be integrated readily by simply counting pulses over the desired period of time. This makes recording the data particularly simple and permits the use of a wide variety of devices for manual reading, automatic printing, and recording on computer-compatible equipment for further processing.

Integration can be carried out by means of a totalizing counter, a printing counter programmed to print at regular intervals, such as 15 minutes or hourly, or by direct recording of the pulses on magnetic tape for subsequent translation, as described in Chapter 1.

In the following section the pulse-gating multiplication technique is discussed in more detail.

PULSE-GATING MULTIPLICATION

DESCRIPTION OF THE METHOD

Measurement of energy flow generally requires that the rate of mass flow of the fluid be multiplied by the temperature difference between inlet and outlet in real time. However, multiplying two variables in real time is usually complicated, because it involves conversion to digital form, multiplication by computer, and recording in digital form on a suitable medium. Alternatively, the two quantities may be recorded on a time-base and then multiplied by computer. Pulse-gating multiplication offers another alternative for multiplying two variable quantities in real time; the result is in a very convenient form that permits the use of a wide variety of inexpensive recording devices for obtaining virtually instantaneous readings of the rate of energy flow or for integrating energy flow over long periods of time.

Principle

The two variable quantities to be multiplied first must be converted to pulse rates of appropriate frequency ranges. The frequency ranges selected for the two variables must be several orders of magnitude apart to satisfy resolution requirements. The relationship between change in the variable and change in the frequency of the pulse rate must be linear. Furthermore, the relationship must provide "zero" frequency at zero value of the variable. This requirement is virtually impossible to achieve in practice but generally can be approached with sufficient proximity to avoid any problems. As an example, consider a requirement to multiply the difference between the outlet and inlet of a water heater by the rate of flow through the appliance. If we arrange to convert a temperature difference of, say, 60°C to a pulse rate of 10 kHz, varying linearly downwards towards zero Hz at zero temperature rise and also arrange for a water flow meter to generate one pulse for each ¼ liter of water drawn, then we have roughly the requirements outlined above for successful pulse-gating multiplication. The maximum frequencies will be about 10 kHz for a temperature difference of 60°C, and 1 Hz (60 pulses per minute) for water drawn at a rate of 15 liters per minute.

Next, we arrange for the *lower* frequency (the water-flow signal) to operate a gate that turns ON for a *fixed period* of time (say, 250 msec.) *each time* a pulse is produced by the water flow meter. This gate effectively multiplies by one when it is ON and by zero when is OFF. When this ON-OFF gating function is applied to the *higher frequency* (the temperature-difference signal or data signal), the result is a series of pulse bursts of 250-msec. duration, whose frequency corresponds to the temperature rises of each ¼-liter increment of flow. In this manner, we have effectively "tagged" each ¼-liter of water with the temperature information contained in the corresponding data pulse; appropriately, each gated sampling period is referred to as a "tag." Figure 20 illustrates the steps required and the appearance of the data produced. It is useful to check the validity of the method by testing the limits; the following circumstances should result in zero data-signal output:

1. Zero temperature difference and any flow
2. Any temperature difference and zero flow
3. Zero temperature difference and zero flow

The energy flow data are in the form of "tags" of pulse bursts at a frequency of about 10 kHz. Recording data in this form presents problems, because of the high resolution and speed that would be required of the recording medium. For convenience, a "divide-by" circuit can be used to bring the data frequency to a range that can be comfortably handled by electro-mechanical counters, event recorders, or slow-moving magnetic tape for unattended recording over long periods of time. A particularly convenient top frequency for recording is 1 Hz, since it permits the use of a wide variety of inexpensive recording equipment.

Resolution of Pulse-Gating Multiplication

Taking as an example a maximum frequency of 10 kHz for temperature

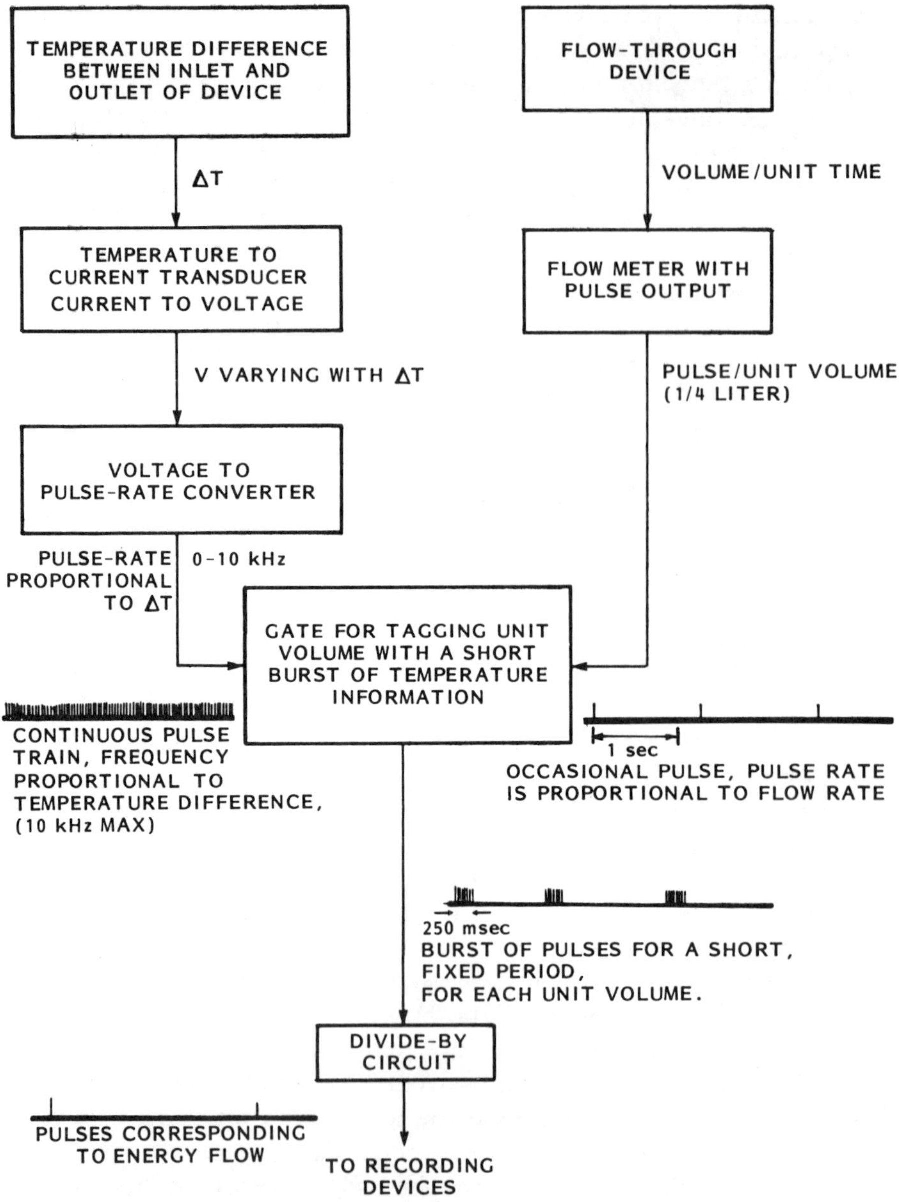

Figure 20. Steps required to multiply temperature difference by flow to obtain energy flow (variable flow).

difference measurement, 1 pulse per ¼ liter of water flow, and a gate-ON duration (tag) of 250 msec., one can determine the resolution of temperature, volume flow, and energy flow under various conditions. Refer again to Figure 20. With a temperature difference, ΔT, of 60°C, producing a pulse rate of 10 kHz, a resolution of ±1 count represents the following temperature difference resolution:

$$\frac{\Delta T}{\text{Total pulses in the tag}} = \frac{60}{10{,}000 \times 0.25} = 0.024°\text{C}.$$

Because of the linear relationship between temperature and pulse rate, the resolution of temperature difference across the entire range remains the same. In general terms, then, the resolution of temperature ΔT_R is:

$$\Delta T_R = \frac{\Delta T_{max}}{F_{max} \times P} \quad (6)$$

where: ΔT_{max}, is the maximum expected temperature difference
F_{max} is the pulse rate at ΔT_{max}
P is the length of the temperature sampling period (tag) in seconds

The resolution of the flow meter is predetermined by the volume-per-output pulse selected; in this case, the resolution is ¼ liter. It should be noted that some meters may have considerable errors at very low or very high rates of flow. Within the range of operation where the error is acceptable, the resolution will be the nominal volume-per-output pulse.

The multiplication is performed on each increment of volume, so that the resolution is 0.024°C per ¼ liter of volume, or, in terms of energy, using equation 4:

$$E_R = V \Delta T_R c \quad (7)$$

Substituting for ΔT_R from equation 6:

$$E_R = V \frac{\Delta T_{max}\, c}{F_{max} \times P} \quad (8)$$

where E_R = Resolution of energy in joules
c = Specific heat capacity (4187 joules/1°C)
ΔT_{max} = maximum expected temperature difference, in degrees (60°C)
F_{max} = Pulse rate at T_{max}, in Hz (10,000 Hz)

V = Increment of volume in liters (¼ liter)
P = Length of temperature sampling period (tag), in seconds (0.25 sec.)

Substituting in equation 8:

$$E_R = \frac{1}{4} \times \frac{60}{10{,}000 \times 0.25} \times 4187 = 25.122 \text{ joules/pulse}$$

For most practical applications, this is an unnecessarily high resolution, and one can afford to reduce it by using a "divide-by" circuit to permit the use of slow-speed recorders without sacrificing the basic resolving capability of the circuitry. In order to end up with a maximum of one energy pulse per second, it is necessary to divide by a factor on the order of the maximum number of pulses expected in a single temperature difference tag, i.e., F_{max}¼P. Since the sampling period P is 250 msec., the maximum number of pulses in a tag will be 10,000 ×0.25 = 2500 pulses. This, then, is the smallest factor that one can divide by in order to end up with, at most, one energy pulse per second. In this instance, one water-flow pulse will represent 25.122 joules × 2500 = 62,805 joules. Converting to kWh by dividing by 3.6×10^6 joules/kWh, 1 water-flow pulse = 0.0174 kWh.

The "divide by" circuit must follow the gating circuit in order to divide energy-pulse data rather than temperature-pulse data. Dividing temperature-pulse data by a large factor is counter-productive, since it reduces the basic resolution of each temperature tag.

Error Analysis

Pulse-gating multiplication involves tagging each incremental water volume with temperature information. Depending on the rate at which each increment of volume is drawn and on the rate of change of temperature difference during the draw, the actual error will vary. The relationship between actual temperature of each incremental volume and the "tag" temperature produced by the instrumentation is illustrated in Figure 21. A constant flow of 15 liters per minute is assumed, along with a linearly decreasing temperature difference of ½ degree per second. Graph A is a plot of volume drawn versus time. Graph B indicates the times (1 second apart) when flow pulses are produced at ¼-liter intervals. Graph C illustrates temperature difference versus time. The temperature tags produced by the instrumentation are represented by

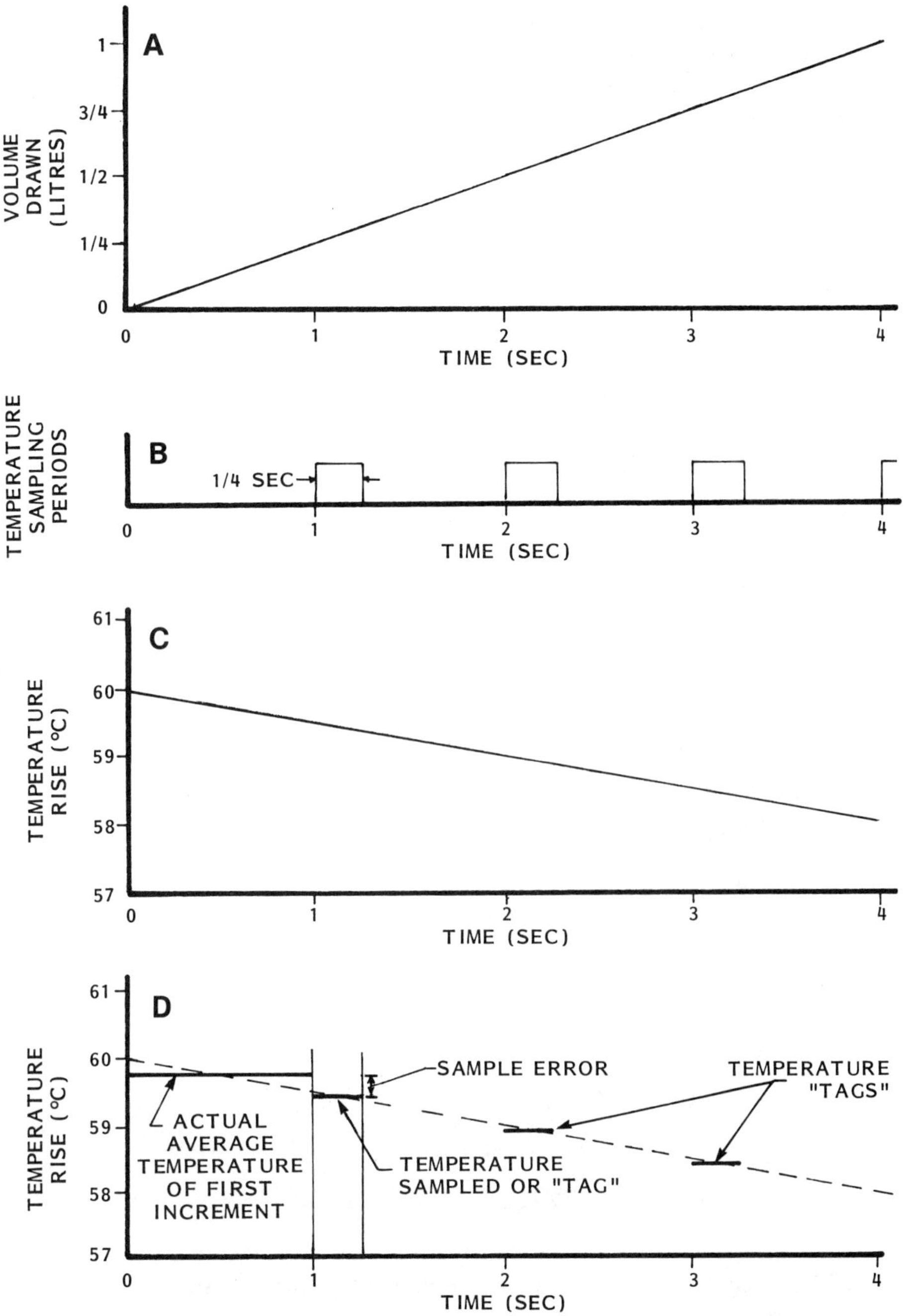

Figure 21. Illustration of sampling error produced by a declining temperature difference.

the solid segments in Graph D. Each temperature tag is associated with the increment of volume that has just been drawn. The sample error can be seen in Graph D and estimated from the slope, sample period, and flow-rate. In the particular case illustrated, the average temperature of the first increment can be calculated from the equation for the straight-line curve, which is:

$$T_D = mt + b$$

where T_D is the temperature difference
 m is the slope of the line ($\Delta T_D/\Delta t$)
 t is time in seconds
 b is the intercept on the vertical axis

Average temperature of the first volume increment, T_1, is

$$T_1 = \tfrac{1}{2}(mt_o + b + mt_1 + b) = \tfrac{1}{2}m(t_o + t_1) + b$$

where $m = \dfrac{-2 \text{ degrees}}{4 \text{ seconds}} = \dfrac{-1}{2}$
 $t_o = 0$ second
 $t_1 = 1$ second
 $b = 60°C$

$$T_1 \text{ (average)} = \tfrac{1}{2} \times -\tfrac{1}{2} \times 1 + 60 = 59.75°C$$

The average temperature of the temperature tag can be quantified by use of the same formula:

$$T_{tag} \text{ (average)} = \tfrac{1}{2}\ m(t_1 + t_2) + b, \text{ where } t_2 \text{ is } 1.25 \text{ sec}$$
$$= \tfrac{1}{2} \times -\tfrac{1}{2} \times (1 + 1.25) + 60 = 59.44°C$$

The sample error in this case is $T_1 - T_{tag} = 0.31°C$.

In general terms, the difference between the actual and the tag temperature is:

$$T - T_{tag} = \tfrac{1}{2}m\ (t_o + t_1) + b - [\tfrac{1}{2}m\ (t_1 + t_2) + b]$$
$$= \tfrac{1}{2}m\ (t_o - t_2)$$
$$= \tfrac{1}{2}m \times (\text{period between flow pulses} + \text{sample period}).$$

The error will be negative with decreasing temperature and positive with increasing temperature. Also, the error will be at minimum during periods of nearly constant temperature difference (slope near zero) and during times of high flows (short periods between flow pulses).

The error will be greatest during times of quickly changing temperatures or very low flows.

In domestic water heating, the rate of change of temperature difference in installations that are adequately sized varies very little because of the relative uniformity of inlet water temperature and the uniformity of temperature within the tank. It should be noted, however, that as water is heated in a tank it expands, backing up into the cold-water supply pipe. This warm water may raise the temperature of the supply pipe and cause a low temperature rise. This can be avoided by arranging to locate the water sensor in a small "bleed" of cold water installed at some point upstream of the meter (see Figure 22). The flow meter must be inhibited from registering pulses when it moves in reverse for any reason, such as thermal expansion of water in the tank. The loss of heat associated with the water backing up into the supply line and being bled down the drain is not registered by this instrumentation. This loss can be calculated together with the stand-by loss from the difference between the measured energy input and energy delivered via the hot water outlet.

CIRCUITRY FOR PULSE-GATING MULTIPLICATION

Conversion of a temperature-difference to a pulse-rate and multiplication of this value by fluid-flow rate, can be implemented in several ways,

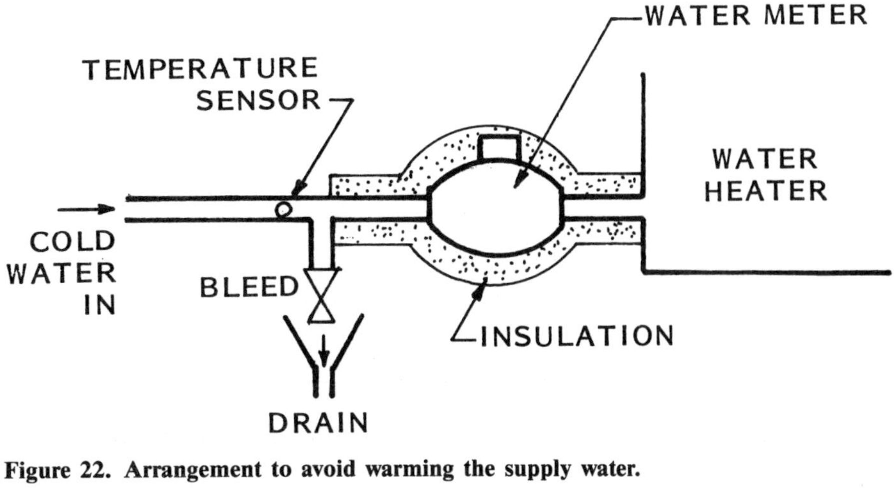

Figure 22. Arrangement to avoid warming the supply water.

all of which are facilitated by the availability of a wide variety of integrated circuits that perform complex tasks and low-cost microprocessors that provide high computing capabilities. The method explained here has the advantages of simplicity and convenience of output form. The flow diagram presented previously in Figure 20 indicates that the two inputs are temperature rise and water flow and the output is in the form of pulses whose frequency corresponds to the amplitude of energy flow. A circuit to carry this out is summarized in the following paragraph.

Temperature difference is sensed by a pair of integrated circuits connected in a bridge arrangement to operate as temperature-to-current transducers. The transducers must be in good thermal contact with the fluid whose temperature is of interest. This can be achieved by equipping the sensors with fins when they are used in an air stream or by immersing them in small wells that are filled with a heat-transfer oil and installed fully into the liquid stream. Probe style sensors can be introduced directly into the fluid streams via appropriate fittings. The voltage associated with the current from the bridge is amplified and changed to a series of pulses by a converter whose characteristic curve essentially intercepts the pulse-rate scale at zero when zero voltage is present (see Figure 23). The pulse train is the data signal.

Fluid flow is sensed by a flow meter whose output is used to trigger a one-shot multivibrator at a rate proportional to the rate of fluid

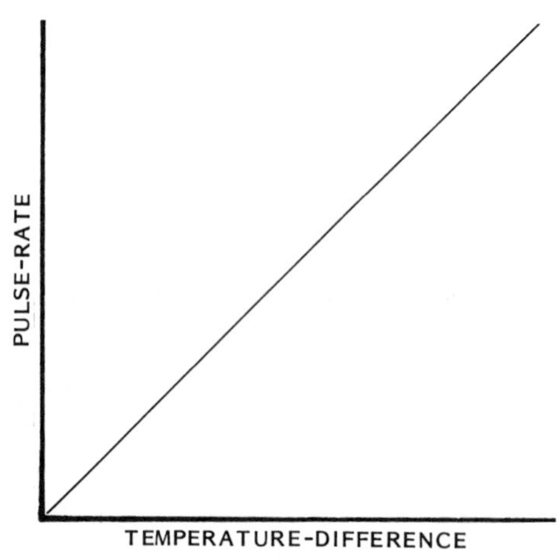

Figure 23. Characteristic curve of temperature difference to pulse rate converter.

flow. Each triggering of this circuit produces one relatively short ON pulse followed by a relatively long quiescent period. This ON pulse is used essentially as a gate that controls the flow of the data signal representing the temperature difference. Each burst of data pulses released by the gating circuit during its ON period is applied to a counter or "divide-by" circuit, which reduces the pulse frequency to the level appropriate for operating electromechanical recording devices (counters, tape recorders, etc.).

Appendix A provides a more detailed description of this circuit and a calibration procedure. The circuit can be built on a single circuit board about 5″×8″. Along with the temperature sensors, flow meter, and a modest power supply, it becomes a most convenient and versatile heat-flow integrator. The circuit has been used successfully as the central instrument for evaluating the performance of a wide variety of heating and cooling appliances.

INDIRECT METHOD

The measurement of energy flow in a system employing forced circulation of the medium (air, water, antifreeze, etc.) can be greatly simplified if the rate of mass flow of the fluid is relatively constant over a day or more while the pump or blower is operating. This section describes a method for establishing the relationship between temperature and energy flow in such special cases.

The method involves injecting a known amount of energy into the fluid stream, mixing the heated fluid thoroughly to ensure a uniform temperature distribution, and measuring the resultant temperature rise of the fluid stream. This is a form of "calibration" that establishes the relationship between power input and temperature rise. Since the measured temperature rise is directly proportional to the rate of heat-energy input, which is known during a calibration, any subsequent temperature rise can be related to the rate of energy input by the heating plant. In this manner, the energy quantification has been reduced to simply measuring temperature rise and relating it to the temperature rise produced by the known heat source. This simple principle is illustrated in the following subsections for a forced-air heating system and for a forced-circulation hydronic space-heating system.

FORCED-AIR SYSTEMS

A typical oil or gas forced-air furnace is illustrated in Figure 24; it is equipped with a supplementary electric heating element (typically 10 to 20 kW) downstream of the combustion chamber, an air mixing device, and two temperature transducers to provide an output signal proportional to temperature rise. The location of the temperature sensors must be selected so that the response is directly related to the bulk air temperature rise. The sensor in the return air stream must be located away from duct walls and away from air inlets that might supply air at a different temperature than that of the bulk. Sensors with response times not longer than a few seconds must be used because very rapid changes in air temperature—on the order of 8°C per minute—normally occur in the supply duct. Mixing of the air must be thorough to ensure a uniform air temperature and thereby permit a single-point temperature measurement to represent the bulk temperature at the location of the temperature sensor. The sensor in the supply air stream must be located near the centre of the duct and away from elbows and outlets. To ensure that the entire contribution of the heat sources appears at the point of measurement, insulation of adequate thickness must be installed on the length of duct between the source and the downstream temperature sensor.

The "calibration" procedure can be as follows:

1. Block the operation of the burner and operate the blower continuously until the indicated temperature difference across the furnace is zero degrees. Because of the long time constant of furnaces with massive combustion chambers, this step may take well over one hour if the furnace has been operating.
2. Energize the electric heating element, wait until the temperature rise stabilizes (30 minutes or more), then measure the temperature rise across the furnace, and note the power input (including that of the blower motor). The "calibration factor" for the furnace, then, is:

$$\text{Calibration Factor} = \frac{\text{Temperature rise at steady state (Celsius)}}{\text{Power input in kW}}$$

$$= \frac{\Delta T_1}{P_1}$$

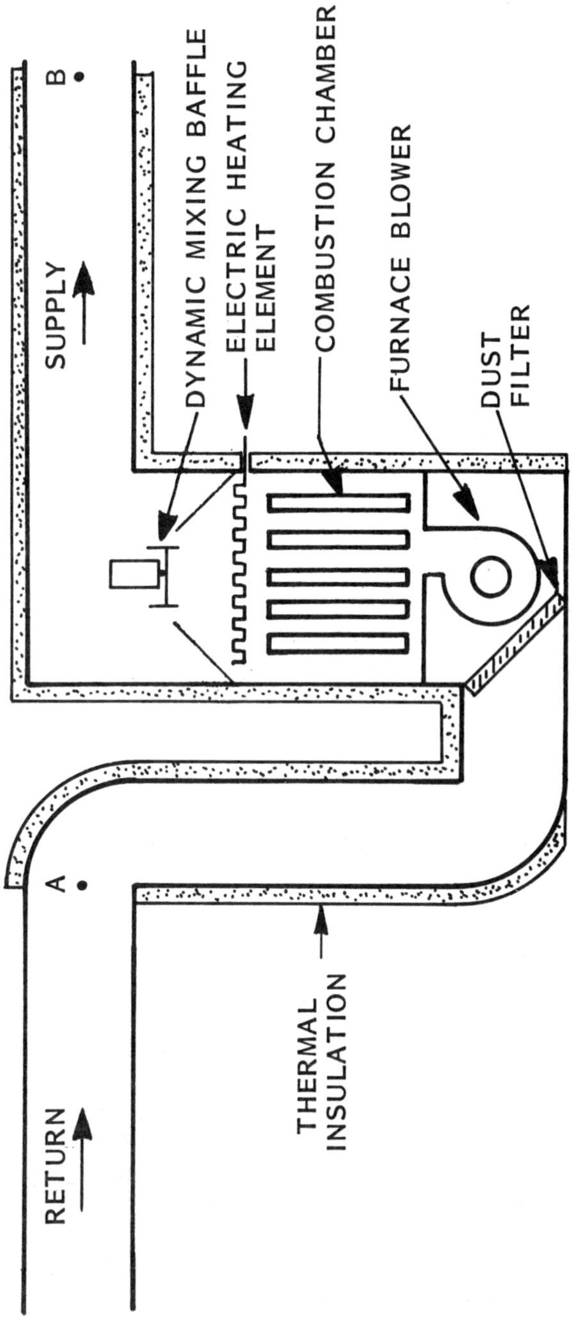

Figure 24. Fuel-fired forced-air furnace with calibration element, mixing baffle, and temperature sensors shown at A and B.

The relationship between power input or rate of heat output at steady state and measured temperature rise is linear and can be illustrated as shown in Figure 25. The slope represents the calibration factor.

3. De-energize the electric heater and restore normal operation of the furnace burner.
4. Record temperature rise during normal operation. Typical records for oil- and gas-fired forced-air furnaces are shown in Figures 26 and 27, respectively.

Since temperature rise has been related to a power output, the Y-axes of Figures 26 and 27 can be replaced by the power equivalent obtained from the calibration factor illustrated in Figure 25. To evaluate energy delivered during the cycle, power output (rate of heat output) must be integrated with respect to time over the duration of the blower cycle or "the area under the curve" (shaded portion of Figures 26 and 27) must be determined.

Since we now have a linear relationship between rate of heat output and temperature rise, all we need is the flow information to permit use of the heat-flow integrator described earlier. First, the temperature-rise signal is converted to a pulse rate in a linear fashion as shown previously in the characteristic curve of Figure 23. Since

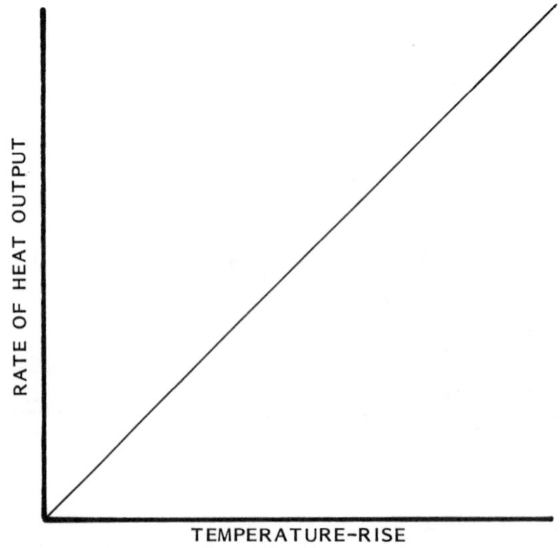

Figure 25. Relationship between rate of furnace heat output and temperature rise.

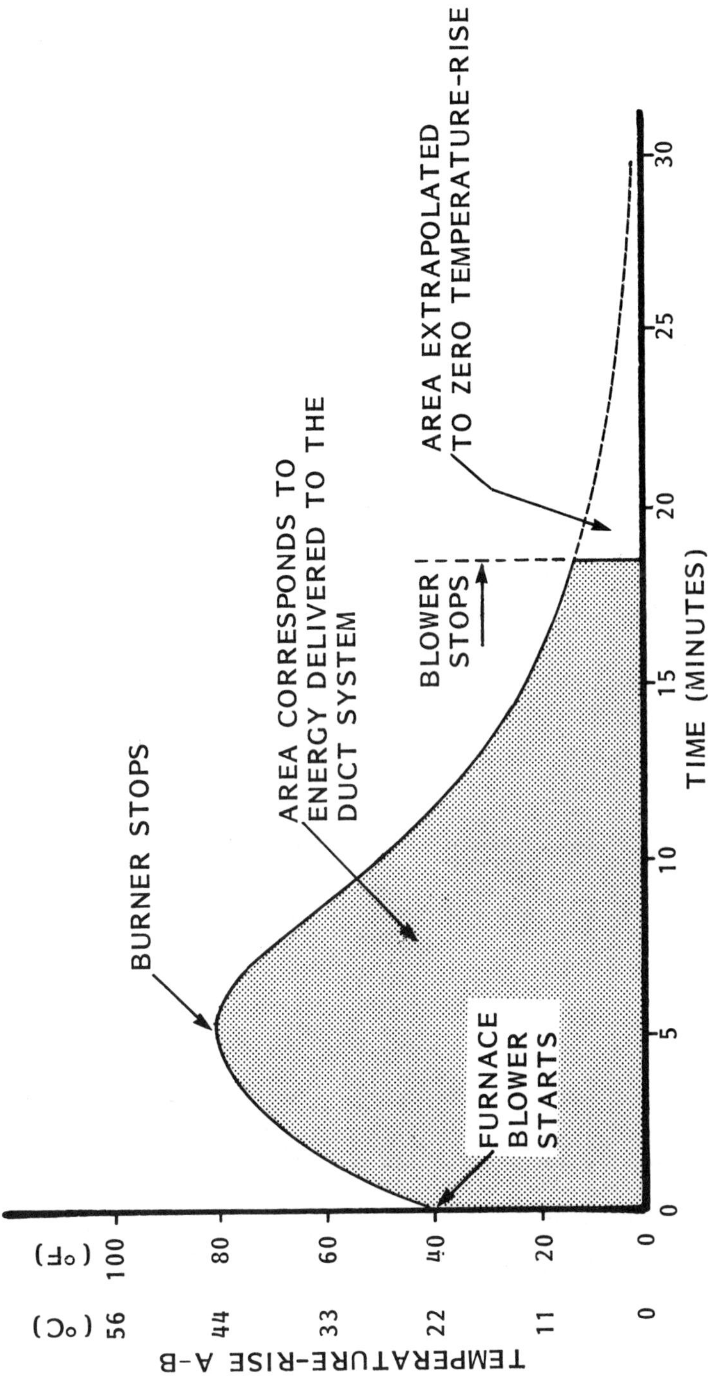

Figure 26. Typical temperature rise across an oil-fired forced-air furnace during a normal cycle.

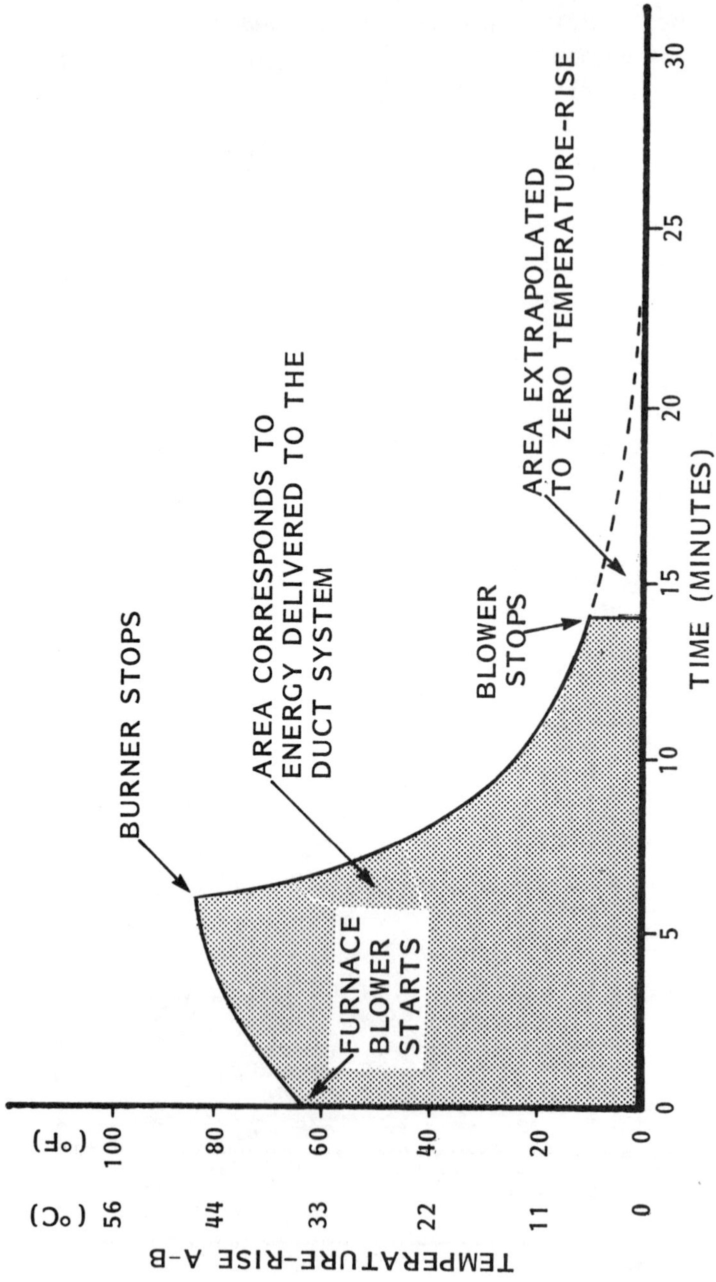

Figure 27. Typical temperature rise across a gas-fired forced air furnace during a normal cycle.

temperature rise is directly proportional to the rate of heat output, as shown on Figure 25, there is a direct and linear relationship between rate of heat output and pulse rate, as illustrated in Figure 28. Then, integration with respect to time becomes a simple matter of counting the total number of pulses during the period of interest. Effectively, we have specified the same conditions that permitted us to use pulse-gating multiplication to multiply temperature rise by flow rate in the previous section. In this case, the time period for integrating energy is selected by gating the pulses to coincide with operation of the furnace blower. When the blower stops, the pulses corresponding to temperature rise are no longer counted. The flow diagram in Figure 29 illustrates the similarity between this method and the previous method for direct measurement of heat output illustrated in Figure 20. In the present case, the method is simpler because the flow is either ON or OFF; when ON, it is constant. Multiplication is performed by a relay that multiplies by one when closed and by zero when open. The energy equivalent of each pulse count can be obtained from a revised "calibration" procedure as follows:

1. Operate the furnace blower continuously until no measurable temperature rise is observed; this will remove all heat that may have remained in storage in the fire-box.

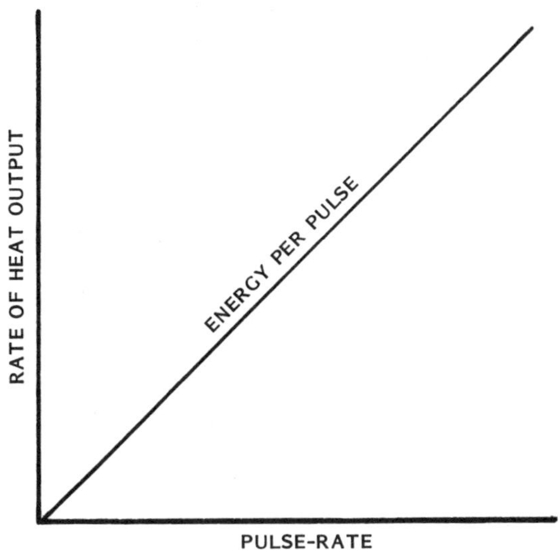

Figure 28. Relationship between rate of heat output and pulse rate.

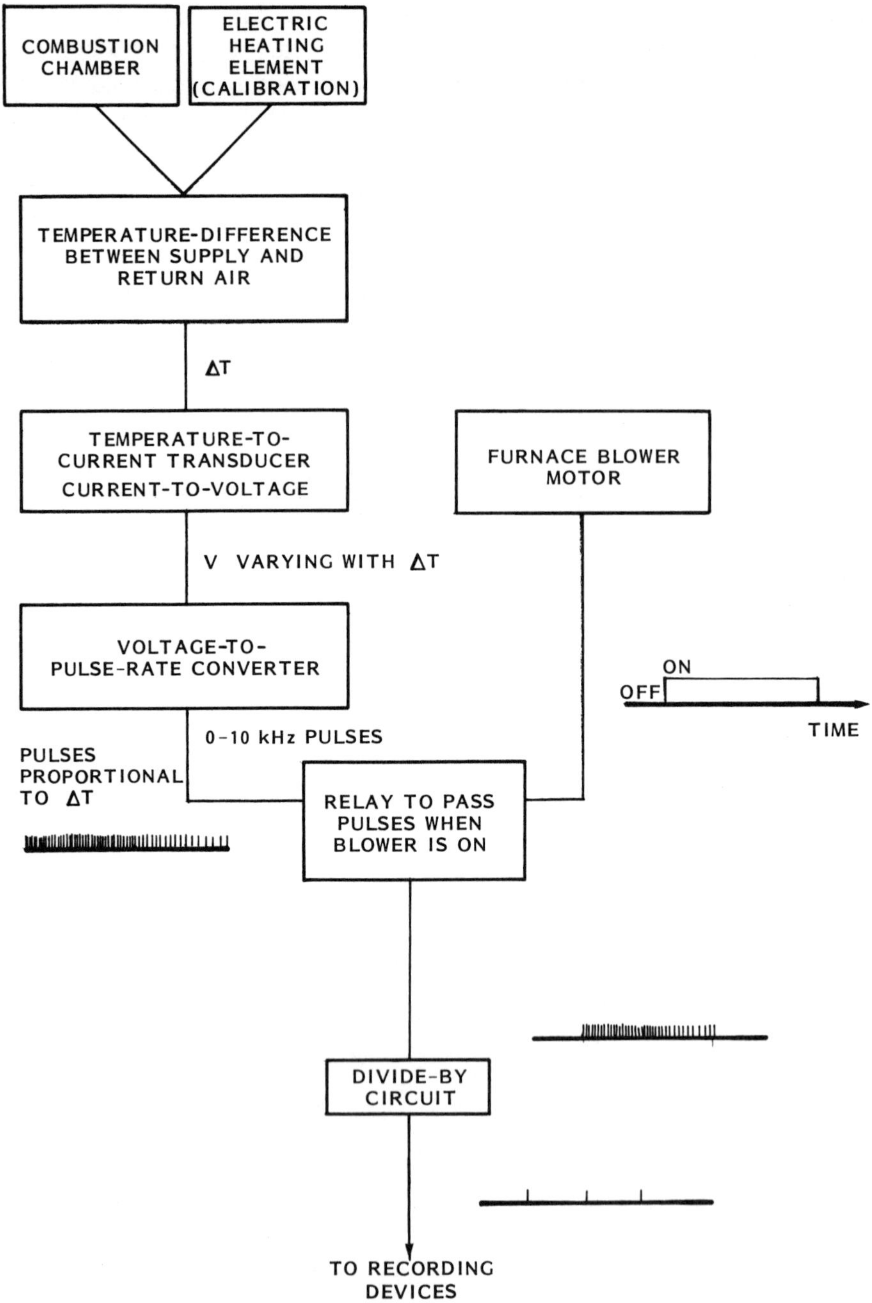

Figure 29. Steps required to multiply temperature difference by flow to obtain energy flow (constant flow).

2. Energize the electric heating element and await steady-state temperature rise (about 30-40 minutes).
3. Count the number of pulses accumulated over a fixed, measured length of time—say, 20 minutes (N_1)—after steady-state has been reached.
4. Note the number of kilowatt-hours (E_1) consumed by the heating element over the same measured length of time.
5. The calibration factor is E_1/N_1, and the units are kilowatt-hours (kWh) per pulse.

As long as the rate of mass flow does not change, the energy delivered by the furnace during any subsequent cycle is simply the total number of pulses during the period of blower operation multiplied by the calibration factor.

A typical plot of temperature rise versus time during a calibration cycle is shown on Figure 30. The energy E_1 noted during the time period of interest, from 40 to 60 minutes in the example, is shown shaded.

Changes in the rate of mass flow through a furnace due to such conditions as clogged air filters, changes in setting of air dampers, changes in supply voltage to the blower motor, and others will affect the value of the calibration factor. Some of the slowly varying changes can be compensated for by carrying out periodic calibrations. Others may be avoided by proper selection of motors or power supplies and design of the experiment. Changes in atmospheric pressure and in return air temperature cause changes in air density. Blower delivery characteristics change slightly with air density and, as a result, a small variation in the rate of air flow may occur during changing atmospheric

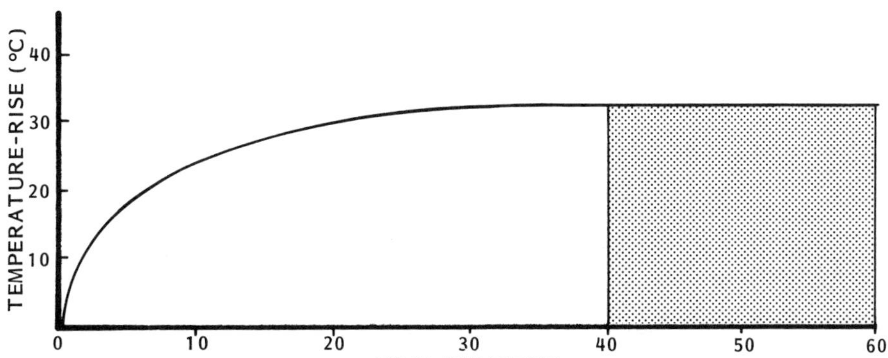

Figure 30. Temperature rise versus time during a calibration.

pressure and during a normal cycle while the return air temperature varies several degrees. Generally, the variation in mass flow through a blower will be about ⅓ of 1 percent per °C change in temperature, depending on the specific characteristics of the rotor. Normally, the return-air temperature of a furnace varies by about 2 or 3°C. In furnaces where the blower motor is thermostatically controlled and operates intermittently, natural convection may take place during the first few minutes of firing prior to the blower operating and also after the blower stops, because some residual heat remains in the heat exchanger. This amount of energy enters the ductwork but is not registered because the instrumentation assumes the air flow to be zero because the blower is not ON. This error can be reduced in various ways: by selecting a low cut-in-temperature setting for the blower thermostat (bonnet control), by arranging for the blower to start at the same time as the burner, or by compensating by calculation. It may be eliminated altogether by operating the blower continuously, if this is acceptable to the occupants of the house.

Other considerations in houses with forced-air heating systems are the amount of energy used by the blower motor and the possible additional energy lost by infiltration during operation of the blower. The heat lost by the blower motor is picked up by the circulating air and would therefore be registered correctly by the instrumentation. However, the motor loses considerably less than half of its energy in the form of heat. The balance is lost in the form of friction in moving air against the resistance of the ductwork and grilles, vibration of metal parts, and other circumstances.

Forced-air systems are leaky; no special efforts are made to make the joints airtight. The return-air plenum is often connected to joist spaces rather than to a well-defined return-air duct system. This practice, along with improper balance of supply- and return-air volumes, leads to the formation of high pressure and low pressure zones in parts of the building, which increases the amount of air leakage to the outside. Longer blower operating periods may lead to higher losses in some buildings, too.

The preferred method is to allow the blower to operate under bonnet control and to perform two tests to obtain a measure of the energy gain by convection before and after a normal blower cycle. One test measures the heat output from the furnace prior to the normal blower ON time, and the other test measures the output after the normal OFF time. This energy can be added as a fixed correction to the energy of each normal cycle of the furnace and totalled with the instrumentation measurement.

As will be seen in later chapters, this method of measuring heat output from forced-air heating plants can be applied to various parts of solar heating systems to determine their performance.

LIQUID SYSTEMS

A method of measuring the heat output of hydronic heating systems can be devised in a parallel manner to the one discussed above when the rate of flow of water is either relatively constant or zero. The method again involves installing an electric heating element downstream of the heat exchanger, measuring the temperature rise produced by the known amount of electric power input, and thus obtaining a calibration factor for the furnace in terms of °C rise per kW of input. The furnace and instrumentation layout is illustrated in Figure 31. The electric immersion element can be installed conveniently in an enlarged portion of the pipe downstream of the heat exchanger. Temperature sensors are required upstream of the furnace and downstream of the immersion heater to provide a signal proportional to temperature rise across the entire heating plant. The temperature sensors must be located in such a manner that all the return fluid and all the supply fluid flows over them. Also, the fluid must be at a uniform temperature at each point of measurement. Taking these precautions permits a single point measurement to be representative of the bulk liquid temperature at each of the sensor locations. Temperature sensors with response times of just a few seconds are required because, during a normal cycle of the heating plant, fluid temperature may change at a rate as high as 8°C per minute. The section of plumbing between the heat exchanger and the downstream temperature sensor should be well insulated to avoid losses that would introduce an error in the measurement of output from the furnace or from the calibration element.

To obtain a measure of energy flow, it is normally necessary to multiply temperature rise by the rate of flow of the heat- transfer fluid. However, in forced-circulation systems where the flow rate is relatively constant over many days, the simpler procedure of using the calibration element and calibration factor can be used as described in the previous subsection. Where changes in flow are expected because of changes in settings of zone controls or other occurrences, this method

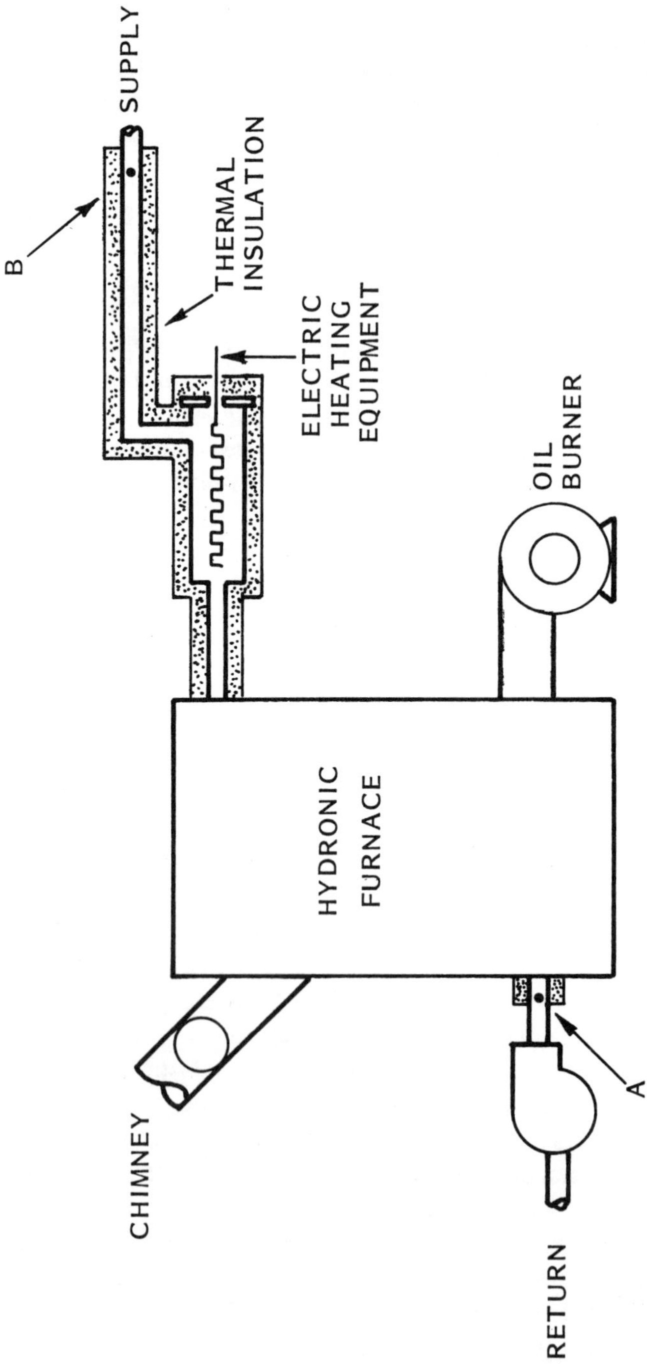

Figure 31. Fuel-fired hydronic furnace with calibration element; location of temperature sensors shown at A and B.

may not be used. The procedure involves blocking the operation of the combustion chamber and allowing it to cool down completely and then energizing the electric heating element until the temperature rise across the heating plant stabilizes. The measured temperature rise divided by the electric power input is the calibration factor. This factor can be used to convert any temperature rise produced by the combustion chamber to an equivalent power output as long as the rate of mass flow through the circulating system remains constant:

$$\text{Furnace output power} = (\text{temperature-rise}) \times (\text{calibration factor})$$

To obtain energy output, the measured power output must be integrated over time, which can be accomplished in the manner described earlier in this chapter. The calibration consists of operating the calibrating element alone and then, after the temperature rise stabilizes, counting the number of pulses over a fixed time period—say, one hour. The increment of energy represented by each pulse then is:

$$\text{Energy per pulse} = \frac{\text{kWh consumed in 1 hour}}{\text{Number of pulses counted in 1 hour}}$$

During normal operation of the heating plant, each of the pulses recorded during the time that the circulating pump operates represents the same increment of energy as was recorded during the calibration.

This energy-flow-measuring system is not suitable for use with liquid systems that rely on natural convection for circulation because the requirement that the rate of mass flow must be zero or constant is not met. A related problem occurs in forced-circulation systems when operation of the circulating pump is controlled by a space-heating thermostat and thereby operates intermittently. In this circumstance, steps must be taken to prevent flow of liquid through the system due to natural convection when the pump is OFF. Otherwise, a significant error will be introduced, because energy will move out of the furnace without its being registered. The problem can be alleviated or eliminated by installing somewhere in the loop a check valve with a small amount of forward pressure, by installing an anti-convection trap such as an inverted "U" downstream of the heating plant, or by operating the circulation pump continuously.

INSTRUMENTATION FOR INDIRECT METHOD

The previous Section described an indirect method for measuring the energy output of heating systems having constant or zero flow rates. It involved obtaining an accurate measure of average bulk temperature difference, converting to a pulse rate, calibration by an electric heating element of known capacity, and multiplication by 1 or 0 corresponding to normal flow or no flow. Integration over time involved simply counting the final energy pulses over the period of interest. The hardware required to perform these tasks is the same as that described for variable-temperature and variable-flow systems except that the gating function is performed by a relay operated by the voltage applied to the pump or blower motor. The relay remains closed and transmits temperature-difference pulses continuously while the pump or blower operates. It is assumed that while the pump or blower is ON the rate of mass flow remains constant and that at other times it is zero.

Energy data in the form of the pulse train produced by this instrumentation can be recorded by a variety of recording media; the choice depends on the way the information will be used. Several such devices are shown on Figure 32. If weekly or daily energy totals will

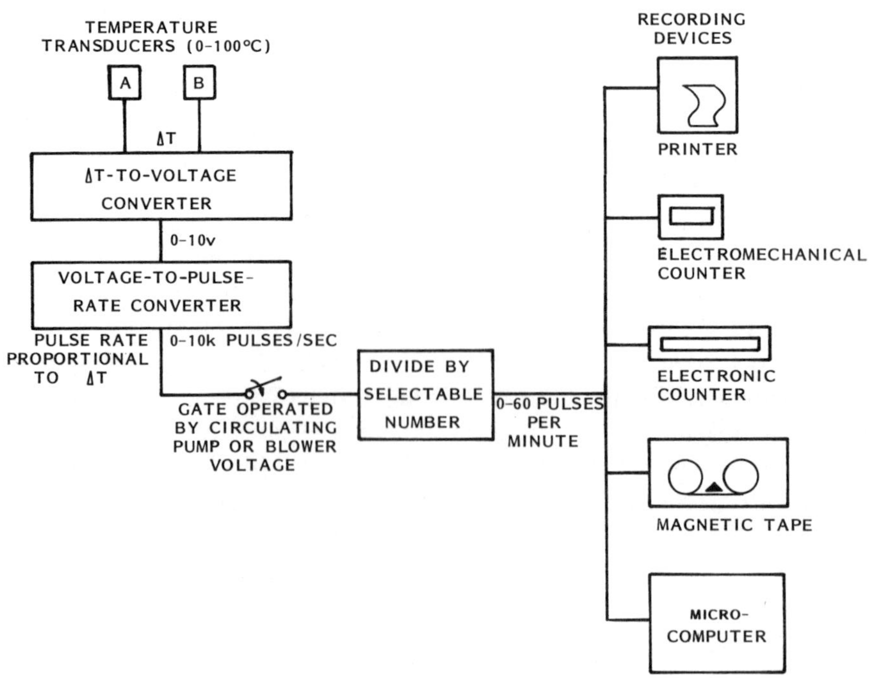

Figure 32. Block diagram of heat flow integrator and various recording devices.

suffice, an electromechanical counter can be used and read manually. If hourly or more frequent readings are required, a printing totalizing counter can be used. This electromechanical device is a combination counter, clock, and programmer, which can be set to print the total counts and time at regular programmable time intervals. Manual processing of the data would be required. For processing the data by computer, a compatible recording medium or transcription of the record into a computer-compatible form would be required. A slow-moving magnetic tape recorder is available for recording pulse data. This type of recorder is used primarily by electrical utilities to record electrical demand. A cartridge of ¼″-wide magnetic recording tape can record over 30 days of information. The pulse data are recorded on one track, while a timing pulse (every 5, 15, 30, or 60 minutes), produced by a clock in the recorder, is recorded on a second track. The data are retrieved by translating the tape onto a computer-compatible magnetic or punched-paper tape using proprietary equipment. The computer can be programmed to produce a listing of total counts recorded between timing marks and to perform the necessary analysis, plotting, etc. Four-channel recorders of the same type are also available, capable of recording three sets of data plus one time track. (See Figures 3 & 4 in Chapter 1 for photographs of this equipment.)

Now coming onto the market are other recording devices that use a microprocessor and solid-state memory to temporarily store the individual totals for each successive time increment. The data are retrieved by interrogation by a microcomputer, which can print them out and transfer them to tape for later processing (see Figures 5, 6, 7, and 8 in Chapter 1). Since counting pulses that correspond to energy is equivalent to integrating energy over time, it is possible to obtain the average power output by dividing the total energy by the period of time over which it was measured. For example, energy data recorded over 15-minute recording periods are in effect average power over 15 minutes. For relatively slowly varying quantities, the records will be very close to the instantaneous actual quantities. This will not be true for intermittent or quickly changing quantities.

3
Applications

The measurement of heat flow in a single pipe or duct may be relatively straightforward. However, in practice, space-conditioning and domestic water-heating systems may not be laid out in a manner that permits the measurement of average bulk temperature of the heated fluid. There may be branch ducts in undesirable areas, or there may be leakage or undesirable flows that make the accurate measurement of total energy flow difficult.

Solar systems generally are more complex than conventional heating systems because of the usual need for a heat-storage element, changes in operating modes, and the greater likelihood of leakage around dampers, ducting, collectors, and heat-storage containers. Depending on where on the system the measurements of energy flow are made, some of the leakage and heat losses may be of no consequence, others may be important. Whenever possible, systems should be designed with the requirements for instrumentation in mind, and provisions should be made for extra lengths of duct, insulation, location of branch ducts, and leak-tight ducting and control devices to facilitate the measurement of energy flows. Depending on the information requirements, relatively simple instrumentation may be adequate—simply metering the fuel input of the system, for example. In other cases, more involved energy-flow instrumentation may be required.

SOLAR DOMESTIC WATER HEATING

MEASUREMENT OF INPUTS

The simplest instrumentation that can be installed on a domestic water-heating system is the type that measures energy inputs—usually a meter to indicate the amount of fuel used and another to indicate the amount of water drawn from the system. In the case of electric resistance heating, an efficiency of 100 percent can be used for the conversion to heat and, by deduction, energy flow and energy use may be determined from the readings of the electric kWh meter and water meter. In the case of solar assisted electric water heating, where the solar component acts as a preheat to the electric tank, water-use data together with the electric-consumption data can be used to calculate the solar contribution. This method is reliable if applied to systems where the solar preheat is provided by one tank and the electric boost by a second tank. In certain one-tank systems, where there is little isolation between the solar and electric components, there can be a problem of electrically produced heat escaping via the collector during periods of little solar contribution. In this case, the amount of heat lost is difficult to estimate, and the solar contribution cannot be estimated with any accuracy.

A reliable measurement of outlet water temperature must be obtained by recording water temperature over at least one day, since the thermostat in a water heater often has an operating differential of several degrees. Stand-by losses of both the solar and the electric tanks must be taken into account, since these represent a significant portion of the energy used. The temperature of the incoming cold water must also be noted periodically, since it may vary significantly with the seasons, depending on the source. For example, river and lake water may vary between 1°C and 22°C with the seasons, whereas well water may be a constant 14°C year-round. The energy flows in the electric-tank system are shown in Figure 33. The energy balance, illustrated in Figure 34 is as follows:

$$H_E = H_W + H_L \qquad (9)$$

where H_E = Electrical energy input (MJ)
H_W = Energy content of hot water (MJ)
H_L = Stand-by loss of the tank (MJ)

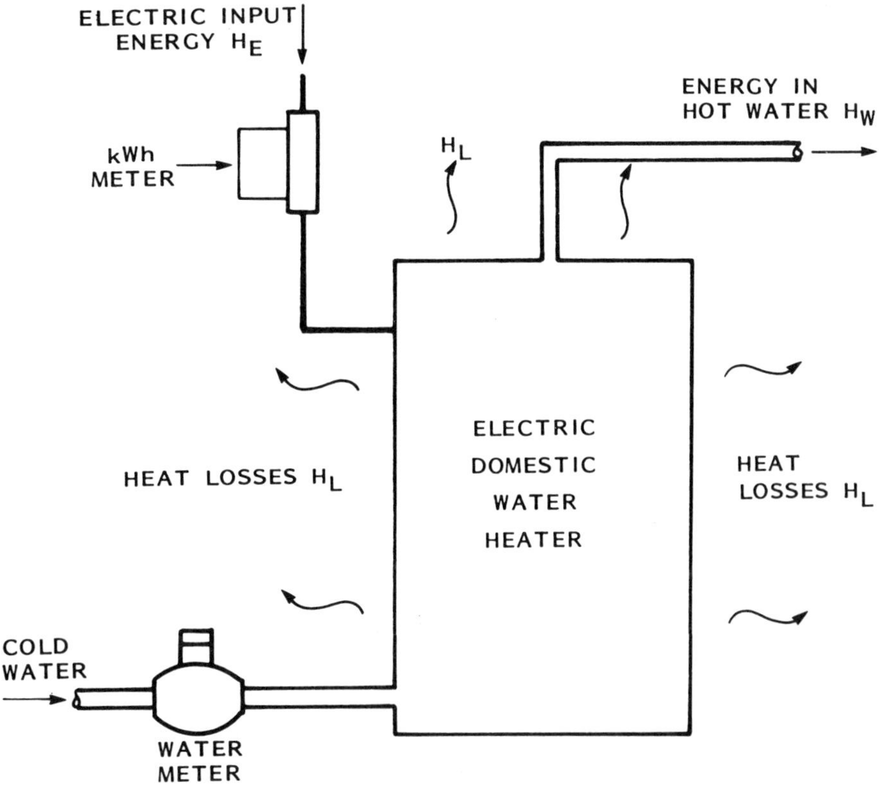

Figure 33. Energy flows in an all-electric domestic water heater.

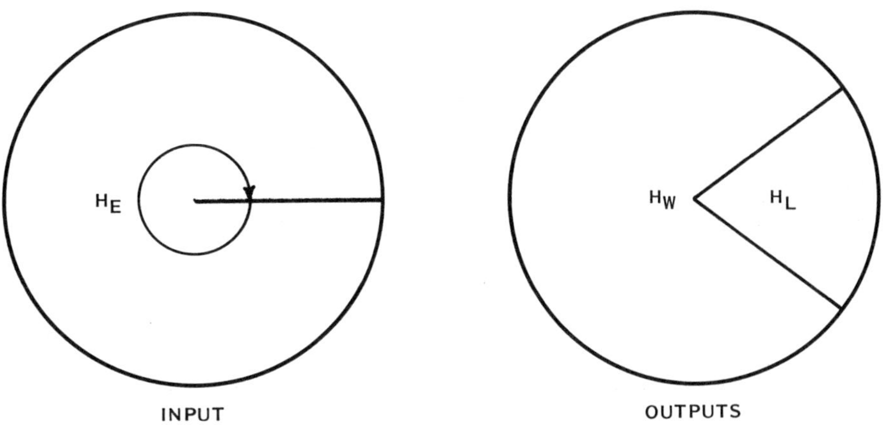

Figure 34. Energy balance in an electric domestic water heater.

The period of observation may be a number of hours, days, or months, as long as the hot and cold water temperatures are known. The energy contained in the hot water drawn is calculated as follows:

$$H_W = Q\ (T_B - T_A)\ c \tag{10}$$

where H_W = Energy content of hot water (MJ)
Q = Volume of water (liters)
T_B = Average outlet water temperature (°C)
T_A = Average inlet water temperature (°C)
c = Specific heat of water (KJ/liter°C) = 4.187

The stand-by loss of the tank may now be evaluated by substituting for H_E and H_W in Equation 9. Alternately, the standby loss may be evaluated in a fully recovered tank by noting the electrical consumption over a 48-hour period with no water being drawn. The losses of an operating tank may, in fact, be somewhat lower, because the tank may be partly recovered over periods of time, reducing the inner tank temperatures and, hence, the loss.

In residential applications, stand-by losses of tanks with 5 cm of glass fibre insulation can easily account for 1/5 of the total energy used by the water heater.

Let us now add a solar preheating system upstream of the electric water heater and analyze the energy flows, assuming that only a water meter and electric meter are to be used. Figure 35 illustrates the system and the energy flows, and Figure 36 illustrates the energy balance that must exist.

The energy inputs to the system are the electrical energy and the solar energy incident on the collector. The energy outputs are: (l) hot water to the plumbing system, (2) stand-by losses of the electric tank (made up by electricity), and (3) heat losses of the solar preheater tank, the plumbing to the electric tank and collector, and the collector (all made up by the solar system). Using the symbols in Figure 35, the heat balance equation is:

$$H_E + H_S = H_W + H_{EL} + H_{SL} \tag{11}$$

where H_E = Electrical energy used over a particular period of time
H_S = Solar energy picked up by the collector over the same period of time
H_W = Energy content of the heated water

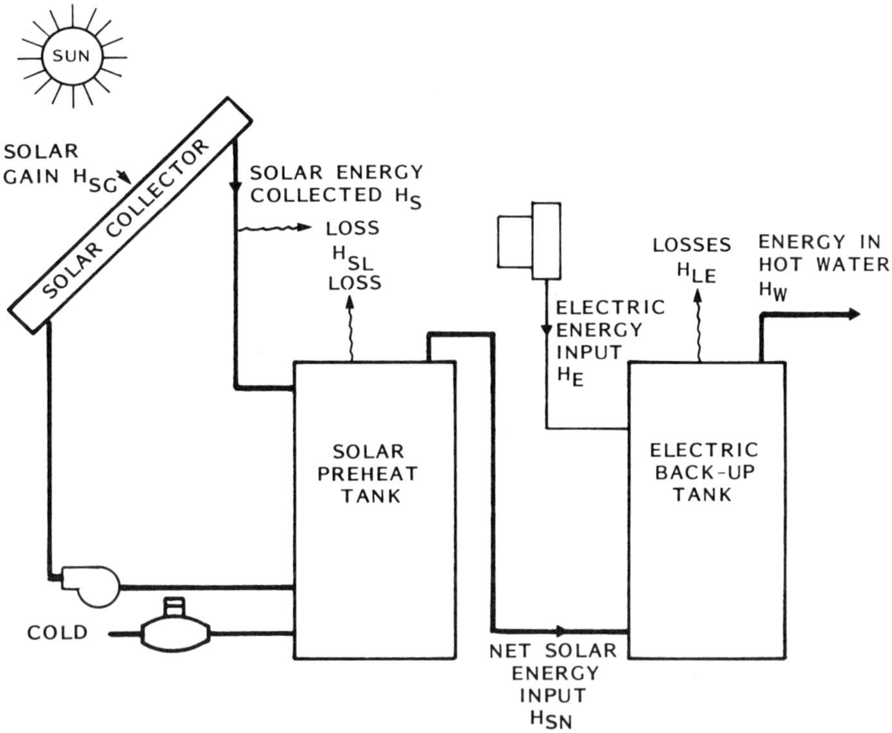

Figure 35. Energy flows in a solar-assisted water heater.

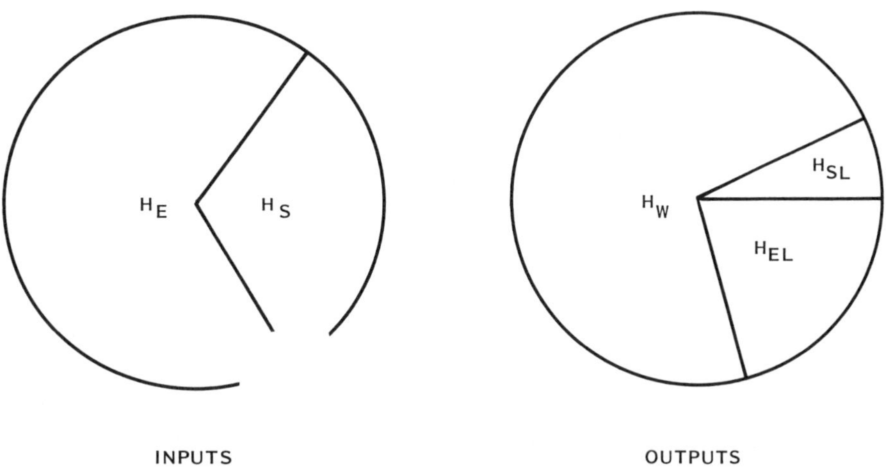

Figure 36. Energy balance in a solar-assisted water heater.

H_{EL} = Heat loss of the electric water heater
H_{SL} = Heat loss of the solar preheater tank, plumbing, and solar collector.

The important quantity to be computed is the net solar contribution, H_{SN}, which is the amount of electrical energy displaced by the solar portion of the system. This may be evaluated by comparing the actual amount of electricity used with that which would have been used by an all-electric system. Suppose that Q_1 liters of hot water were drawn from the system over a given period of time. According to Equation 10, the energy content of the water delivered is:

$$H_W = Q_1(T_B - T_A)\ c$$

Using Equation 9 and substituting for H_W, the electrical energy required by an all-electric system is:

$$H_E' = Q_1(T_B - T_A)\ c + H_{LE}'$$

where H_{LE}' is the quantity of heat loss on stand-by over the same period. Now, the difference in electrical consumption between the all-electric and solar-assisted water heaters is the net solar contribution H_{SN}:

$$H_{SN} = H_E' - H_E'' = Q_1(T_B - T_A)\ c\ + H_{LE}' - H_E'' \qquad (12)$$

where H_E'' is the measured electric energy used by the solar-assisted water heater over the same period.

Equation 12 is in terms of quantities that can be evaluated from the following data:

- Volume of water drawn over the period of interest (Q_1)
- Temperature difference across the entire system from outlet to cold water inlet ($T_B - T_A$)
- Measured or calculated heat loss of the electric tank (H_{LE}')
- Measured electrical consumption of the solar-assisted system (H_E'')

From relatively simple instrumentation and calculation it is possible to obtain a reasonably accurate measure of solar system performance.

A measure of the overall solar utilization efficiency may be obtained if the incident solar energy is recorded. A solarimeter and recording package as described in Chapter 1 may be used to record total radiant energy density incident on the collector surface. By multiplying the energy density by the net collector area intercepted, the gross solar energy input to the collector, H_{SG}, can be determined. The net utilization efficiency will then be:

$$\text{Net Solar Utilization Efficiency} = \frac{H_{SN} \times 100}{H_{SG}}$$

If a fuel-fired water heater is used as the back-up for a solar system, then the calculation of solar system utilization efficiency involves taking into account the combustion efficiency and standby loss of the fuel-fired water heater. Combustion efficiency as measured from flue-gas analysis is usually not representative of the cycle efficiency because of changing temperatures in the walls of the combustion chamber and varying amounts of air for combustion as the flue and flue gases heat up. Standby losses in water heaters with an internal flue are usually much higher than in electric water heaters and vary somewhat with the amount of hot water used. Unless the efficiency and standby loss are known with confidence, it is recommended that fuel-fired water heaters not be used to evaluate the performance of a solar preheat system when only energy inputs are recorded. It is preferable to replace the fuel-fired system temporarily with an electric water heater that has a well defined efficiency and standby heat loss.

In the next subsection, instrumentation to measure energy outputs will be described; these permit the performance of system components to be evaluated directly over short or long periods.

MEASUREMENT OF OUTPUTS

The instrumentation required to measure energy output from a section of a solar water-heating system is more complex than the instrumentation required to measure energy inputs to the system. The evaluation involves the multiplication of temperature difference between fluids entering and leaving the system by the rate of flow of the fluid. Energy flow can be measured at various boundaries in a system, as illustrated in Figure 37, depending on the quantities of interest. For example, energy crossing boundary BB′ is equivalent to the net energy collected by the solar panel. Energy crossing EE′ is the energy contribution of the solar component. Energy inputs are solar energy across AA′ and electrical energy across FF′. Note that heat losses take place from the piping and from the tanks (these are not illustrated). Depending on the boundary selected, these losses may be included or

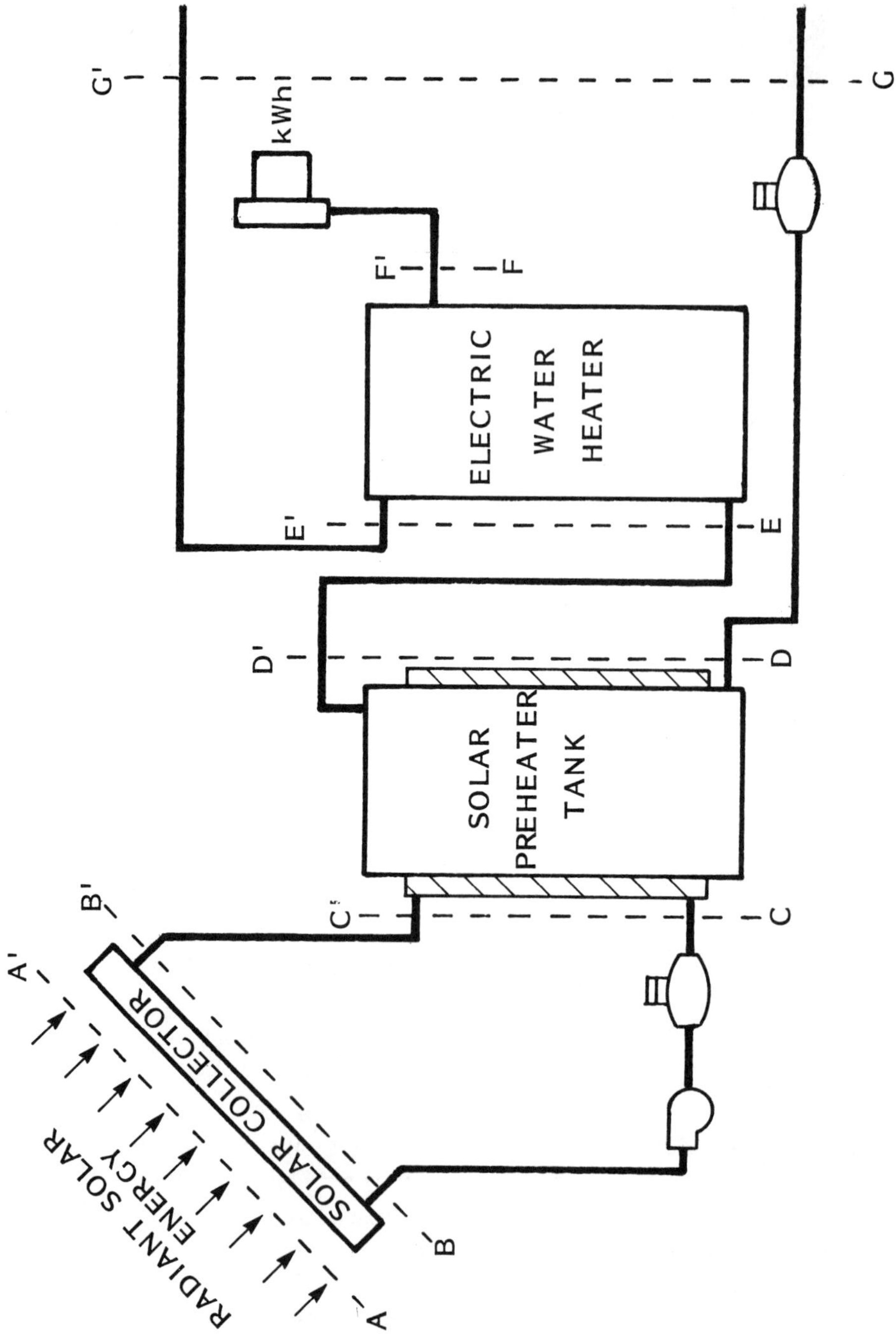

Figure 37. Solar-assisted electric water heater system showing boundaries across which energy flow measurements can be made.

excluded in the measurements. For example, the difference between energy measured across BB′ and across CC′ will be the heat lost by the piping between the collector and the preheat tank. A detailed accounting of hypothetical energy inputs, energy flows, and losses is presented on Figure 38. The piping and tank losses have been obtained by subtraction of energy flows across appropriate boundaries. It is interesting to note that a considerable amount of solar energy is lost to the building by the piping and the solar tank and that a certain amount of electric heat is also lost by the electric tank. The best way to reduce losses is to keep the lengths of pipes to a minimum and to insulate pipes and tanks to high levels.

Measuring the heat output of a solar panel is an instructive example of the application of heat flow instrumentation. The pulse multiplication technique described in Chapter 2 under "Liquid Systems" will be applied to the solar collector loop to permit detailed observation of collector performance. This application may be of interest to solar-system and component designers who require instantaneous information on performance while altering design parameters, such as fluid temperature, flow-rates, glazing angles, and absorber surface.

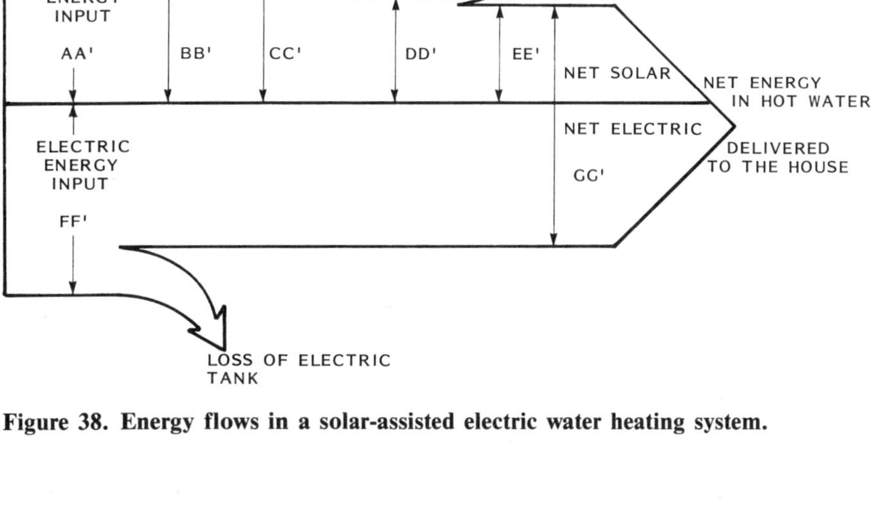

Figure 38. Energy flows in a solar-assisted electric water heating system.

The instrumentation package shown in Figure 39 is required. It consists of temperature sensors A and B installed in the supply and return lines respectively, a liquid flowmeter L equipped with a pulsing head, and the electronic package E described in Chapter 2 to carry out the multiplication and integration. Changes in flow rate caused by variations in viscosity of the fluid with temperature are of little consequence to the performance of the instrumentation. Fluid must not be permitted to circulate by natural convection during sunless periods because the consequent heat loss from storage will not be properly registered by this instrumentation scheme. The specific heat of the heat transfer fluid must be obtained at its mean operating temperature.

The instrument package produces pulses proportional to energy collected. By summing these pulses for 15-minute periods, one obtains the integrated energy over the periods. Sample collector outputs are plotted *versus* time of day in Figures 40, 41, and 42 (broken line) for clear, partly cloudy, and overcast days, respectively.[11]

The radiant energy flux measured by a solarimeter, as described in Chapter 1, may be multiplied by the solar panel's collector area to yield a record of the incident solar energy on the panel. If the

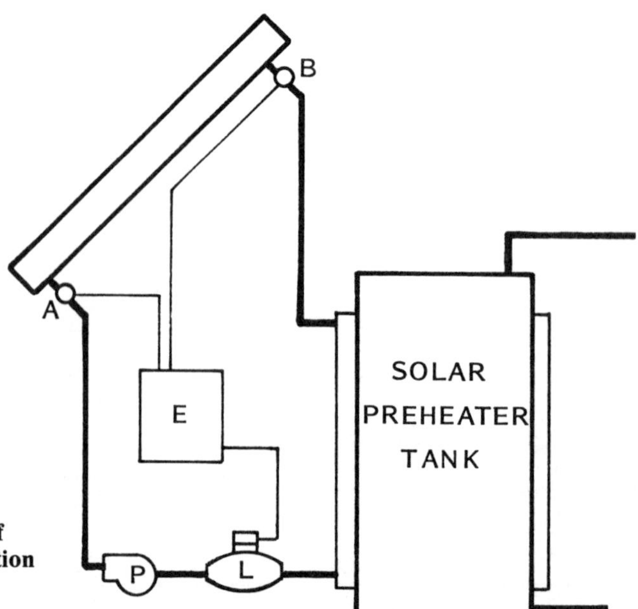

Figure 39. Location of heat-flow instrumentation to measure collector performance.

Figures on this page reprinted with permission from the American Society of Heating, Refrigerating and Air Conditioning Engineers, Atlanta, GA.

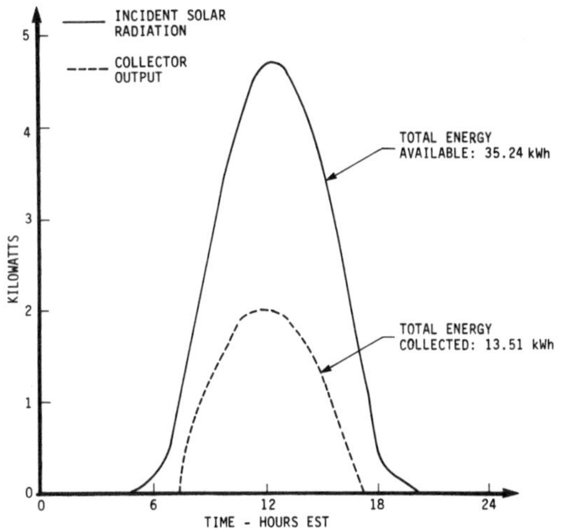

Figure 40. Record of incident solar radiation on a collector and output heat energy of the collector on June 19, 1979, in Toronto—a clear day.

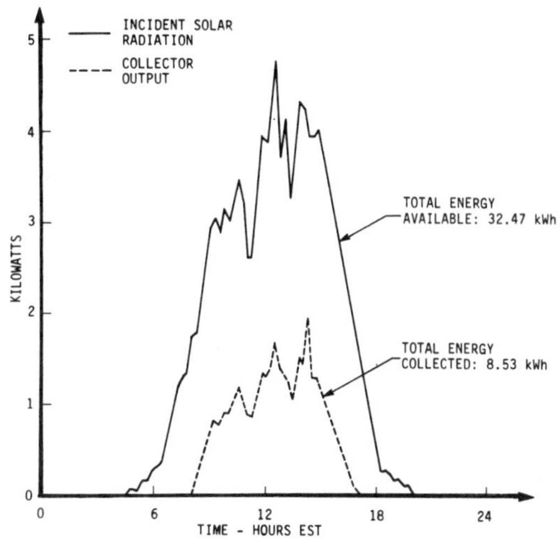

Figure 41. Record of incident solar radiation on a collector and output heat energy of the collector on June 24, 1979, in Toronto—a partly cloudy day.

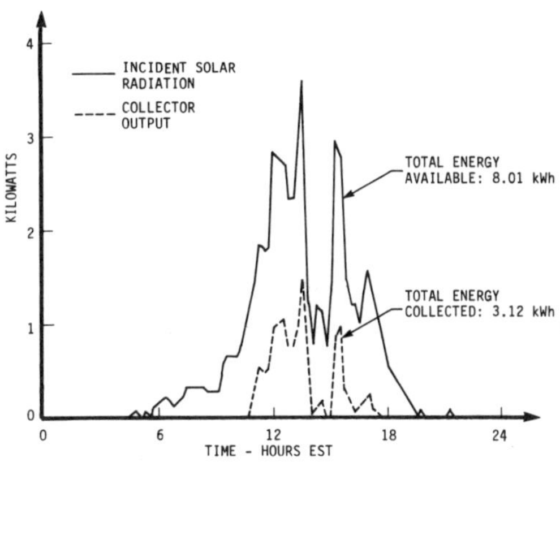

Figure 42. Record of incident solar radiation on a collector, and output heat energy of the collector on June 31, 1979, in Toronto—an overcast day.

solarimeter voltage is converted to a pulse rate and recorded by totalizing the counts every 15-minutes, then graphs of solar energy available during the same clear, partly cloudy, and overcast days can be plotted as depicted by the solid lines in Figures 40, 41, and 42. These graphs give an indication of collector performance under varying insolation conditions.

By totalizing the counts recorded over the day, one obtains the total energy available and total energy collected. The ratio (total energy collected/total energy available) gives a measure of the efficiency on an integrated basis (see Table 5). Calculating the same ratio on a point-by-point basis gives "instantaneous" collector efficiency.

Other factors that must be taken into consideration while assessing collector efficiency include fluid and ambient temperatures, wind speed and direction, and collector fluid velocity.

The experimenter who uses the foregoing procedure can use the results immediately; because the instrumentation pre-processes the temperature and flow information, simply summing pulses results in a quantification of energy. This technique is particularly useful in situations where system components and parameters are being optimized or where the sensitivity of the system to changes in operating parameters is being investigated.

As a second example, let us instrument the system shown in Figure 37 to measure the net energy delivered across boundary GG′. In this

TABLE 5
Total Solar Energy Collected and Heat Energy Available
(From Figures 40, 41, and 42)

Conditions	Energy Collected (kWh)	Energy Available (kWh)	Ratio
Sunny	13.51	35.24	0.383
Partly Cloudy	8.53	32.47	0.263
Overcast	3.12	8.01	0.390
Noon, Clear Day	2.0	4.7	0.42

case, the temperature sensors are located at C and D (see Figure 43) and the flowmeter on the cold water inlet is equipped with a pulsing head. The three signals are processed by the electronics package E, and the output is recorded by an appropriate recorder.

The measurement of H_W can be used with Equation 11 to quantify the net solar contribution:

re-arranging where

$$H_E + H_S = H_W + H_{LE} + H_{LS} \quad (11)$$

$$H_S - H_{LS} = H_W + H_{LE} - H_E$$

$$H_S - H_{LS} = H_{SN}, \text{ the net solar contribution}$$

$$\therefore H_{SN} = H_W + H_{LE} - H_E \quad (13)$$

H_{SN} is now described in terms of the recorded energy H_W, stand-by loss of the electric tank, and electric energy used over the same period.

The solar water heater output, totalized over 24 hours, can be displayed for an entire month as shown in Figures 44 and 45. The lower graphs show the daily performance profiles of the water-heating system

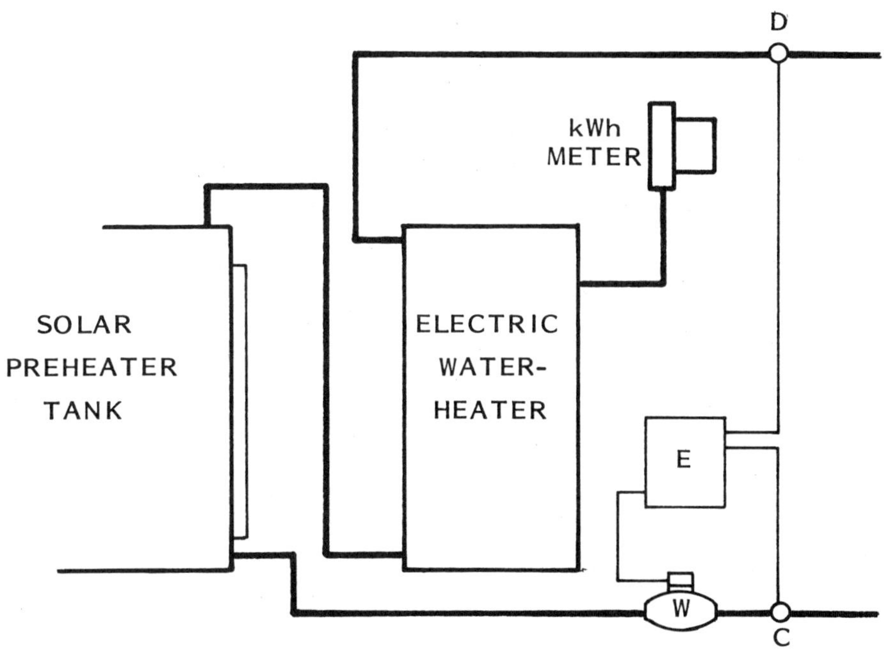

Figure 43. Instrumentation to measure the net energy output of the water heating system.

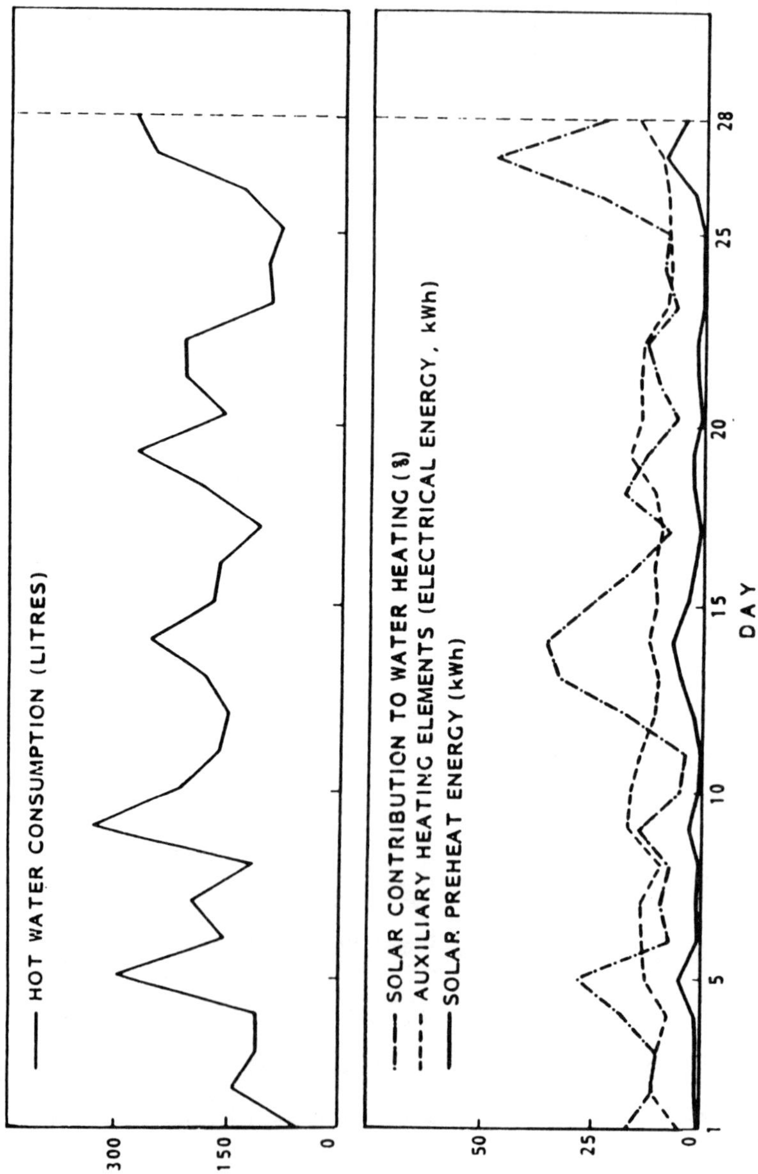

Figure 44. Daily performance profile for a water heater during February, 1981.

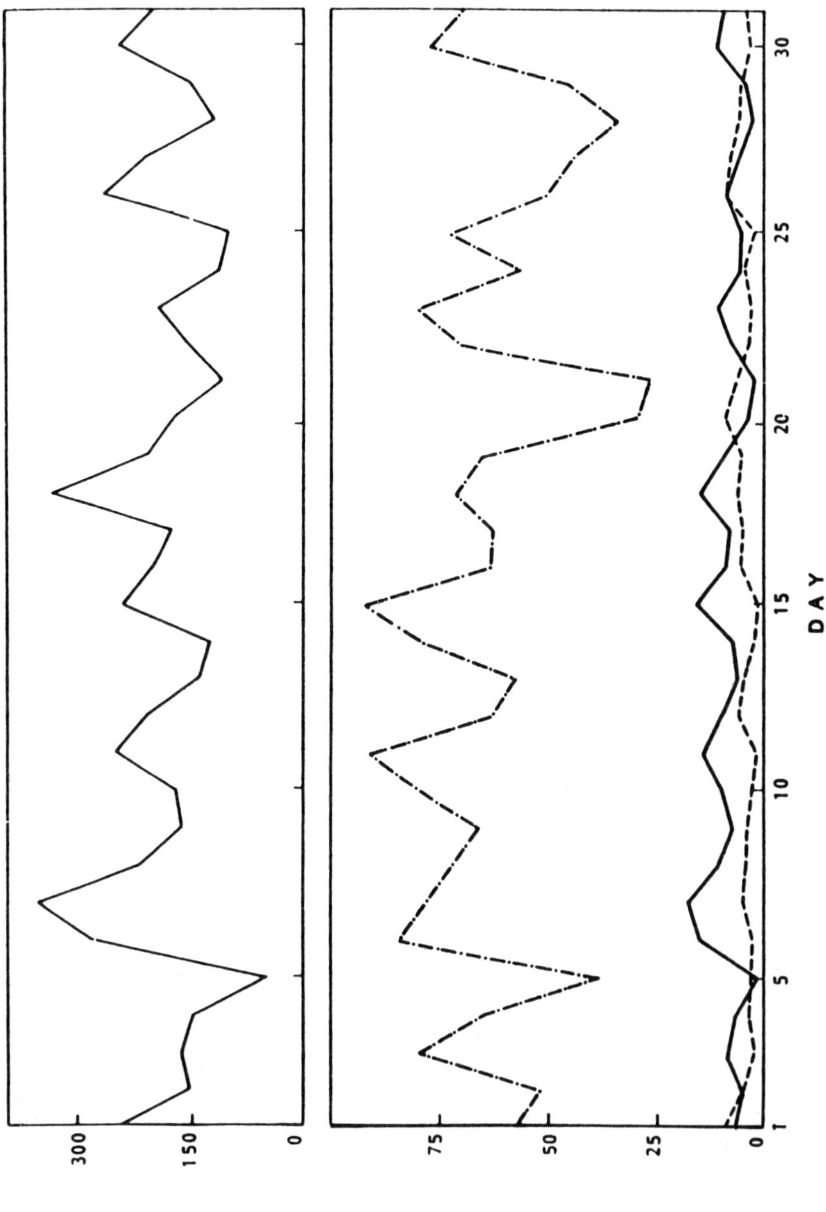

Figure 45. Daily performance profile for a water heater during July, 1981.

during the winter and during the summer, respectively. The heat output measured by the heat flow integration includes the electric back-up energy but not the losses of the electric tank. The net daily solar contribution is obtained by substituting into Equation 13 the measured values of H_W, H_E, and H_{LE}. This quantity is plotted in a solid line along with the electric back-up energy (dashed line) and the solar contribution expressed as a percent (dot-dash line) for each day of the month.

Performance is not only affected by outdoor temperature but also by the amount of hot water used, as illustrated in the upper graphs of Figures 44 and 45.

SOLAR SPACE HEATING

MEASUREMENT OF INPUTS

Following a parallel approach to that used for solar-assisted water heaters, one would expect that an evaluation of performance of a solar-assisted space-heating system could be performed by measuring the total energy input to the building except solar and subtracting this quantity from the total energy requirement to obtain the solar component. Although there is no problem in measuring a building's total energy inputs (gas, oil, electricity, etc.) that contribute toward space-heating, there is a problem in determining the total amount of heat that is required for space-heating. Methods do exist for calculating the heat loss through exterior walls, ceilings, basement wall, windows, floors, and air leakage to assist in selecting the size of heating equipment required. However, when these calculations are used to predict the energy requirement for a certain period of time, such as an entire heating season, the prediction is generally high by 30 to 40 percent.

Correction factors have been developed according to experience; for electric heating a "C" factor, which would normally be 24 (representing the hours per day in the annual-consumption formula) has been evaluated from experience to be 15 to 17.[12]

Another approach that might be considered is to compare the consumption of energy for space heating before and after the solar assist is installed and operated. Problems to consider include the variation of heating degree days from one year to the next (easily 5 percent), variation in wind speed and direction from year to year, and variation in the amount of solar gain through windows due to variation in the hours of sunshine from year to year. Variations in thermostat settings, living habits, age of occupants, condition of weather stripping on doors and windows, shrinkage of lumber, etc., can also have a significant effect on the amount of energy required for space heating. The variations from year to year may be so large as to render untenable the assumption that heating requirements are constant from year to year.

A third approach that may be considered is the use of two identical buildings for a comparison of heating system performance, one with a solar assist, the other without. The heating requirements of two identical buildings heated to the same temperature may not necessarily be the same, since it is virtually impossible to achieve the same degree of airtightness and workmanship in two buildings. It may be possible to adjust the heat losses in order to make them similar, but probably not likely under all conditions.

The use of semi-detached dwelling units may not be suitable, since one will generally be upwind of the other, and the heating consumption will be quite different in each half. Uninsulated party walls, different thermostat settings, and differences in orientation and solar gain of windows can lead to variations in heating requirements of each half of the building by as much as a factor of two.

To add to the complexity of the problem of assessing the energy requirements of a building where gas- or oil-fired furnaces and electric heat pumps are used, the net heat output cannot be determined readily from a measure of energy inputs alone, since efficiencies do vary, depending on a number of factors, such as combustion air setting, outdoor temperature, percent ON time, etc. A measure of the instantaneous efficiency of a furnace can be obtained by measuring the combustion efficiency. The stack-gas temperature, carbon dioxide content, and carbon monoxide, oxygen, and smoke opacity are required. Even then, resulting efficiency is not necessarily representative of cycle or seasonal efficiency, since the combustion conditions vary during the cycle and during the heating season. The assessment of efficiency of a heat pump is complicated by additional factors: the efficiency varies with outdoor temperature and humidity; at low outdoor temperature it delivers less heat than during mild weather; frosting of the outdoor coil affects efficiency—as the ice builds, the air flow through the outdoor unit drops and so does the coil temperature, reducing the efficiency of the device; defrosting the outdoor coil requires heat, degrading the overall efficiency further. Although an estimate of heat pump seasonal efficiency can be obtained from the characteristics of the device and weather records, accurate determinations are very difficult because of the effects of frost formation and defrost cycles on the overall efficiency.

In conclusion, the use of energy inputs to a building alone to estimate the performance of the solar component of a heating system is not reliable, because the total heating energy requirement cannot be accurately estimated.

The performance of solar and other heating systems must be evaluated by measuring useful heat output directly.

MEASUREMENT OF OUTPUTS

Measurement of performance of space-heating systems with solar assist requires the evaluation of fluid flow and temperature rise, multiplication of one quantity by the other to obtain rate of heat output, and integration over the period of time of interest to obtain a measure of the energy delivered. The measurement is to be carried out at various sections of a heating system, depending on which energy flow is required. For example, Figure 46 represents a solar space-heating system with electric back-up. The circulating fluid is a liquid and is stored in an underground tank. There is a liquid-to-air heat exchanger in the furnace and an electric heating element downstream of the heat exchanger. Solar energy impinging on the collector crosses the plane AA′. If the net solar energy collected is of interest, instrumentation would be required across section BB′ to measure temperature rise across the collector and the rate of mass flow through it. It may be that the net useful heat output of the storage unit is of interest. This quantity will not include the heat loss that normally takes place from the piping leading to the collector and from the storage unit itself. In this case, heat flow across section DD″ is required. The net output of the solar heating system into the distribution system can be measured across section EE″. The total heat output to the house can be evaluated across FF″, with the exception of heat loss of the tank, ducting, and piping. These components are shown in the energy flow diagram in Figure 47, and they do contribute to heating the building.

Not all systems lend themselves to measurement of energy flows

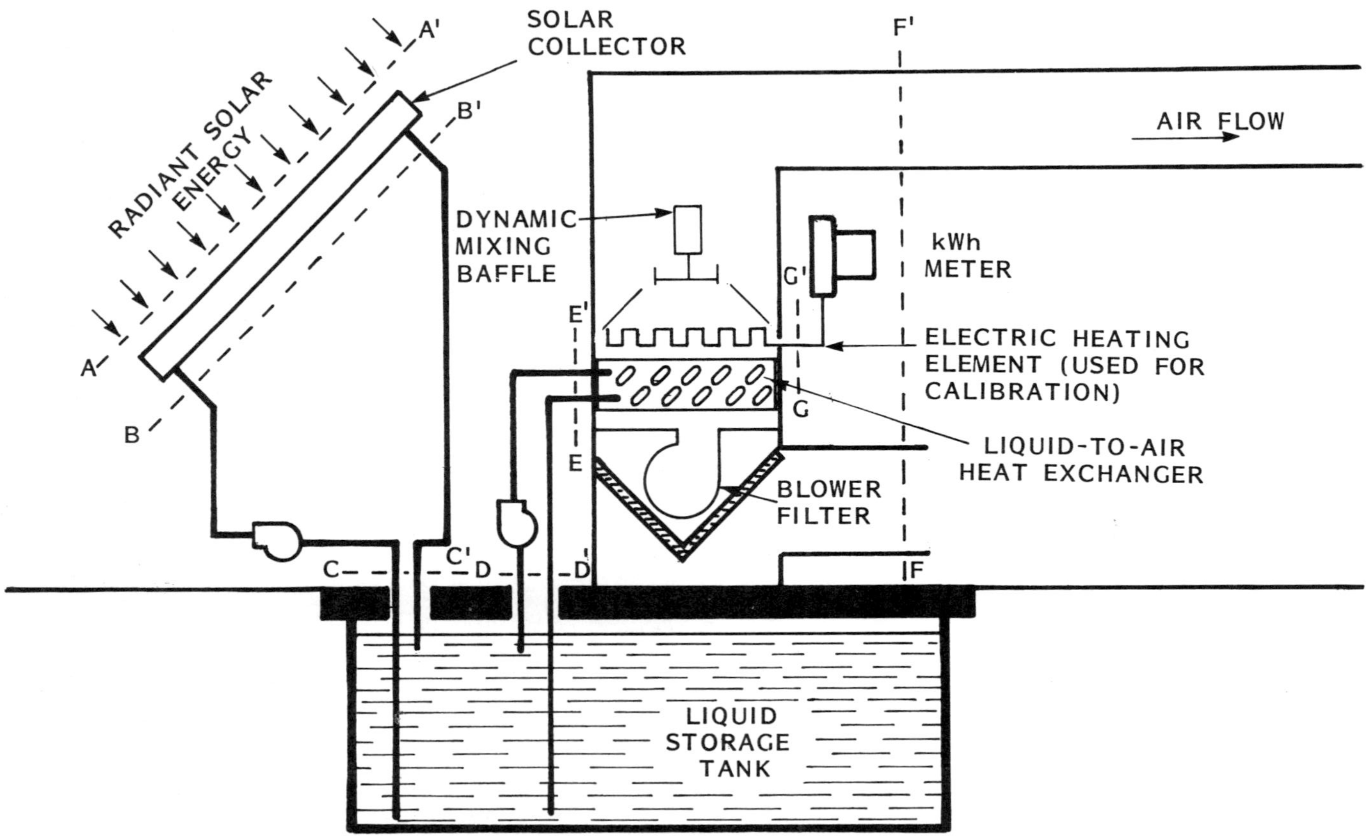

Figure 46. Solar-assisted space heating system showing boundaries across which energy flow measurements can be made.

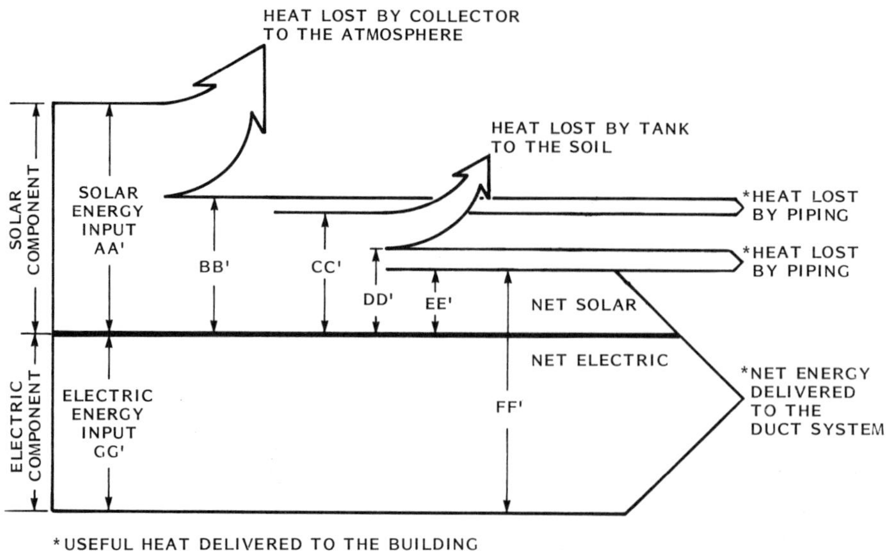

Figure 47. Energy flows in a solar space heating system with in-ground storage.

at all the locations indicated. For example, air-based systems comprised of air collectors, rock storage, and forced-air back-up furnaces have the problem that the rate of air flow may change drastically from one mode of operation to another—from charging the storage to drawing heat from storage, for example. The air flow through the air collector will probably be different when heat is being drawn from storage and when it is not being drawn, because of the additional pressure drop across the storage medium or because of additional pressure induced by the circulating blower. Keep in mind that the product of temperature rise and air flow is required to evaluate energy flow; the assumption used in Chapter 2 for the "indirect method" of measurement—that the air flow is constant—is not vaild here, and other methods must be used. These are discussed in detail in the next section.

Solar systems with isolated collector-storage loop and storage-back-up furnace loop or with all-liquid heat transfer systems do lend themselves readily to energy flow instrumentation. In Figure 48, three sets of instruments are illustrated to measure the heat outputs of the collector, the storage unit, and the heating system into the distribution system. In the following subsections, the instrumentation requirements for air and liquid systems will be described in more detail.

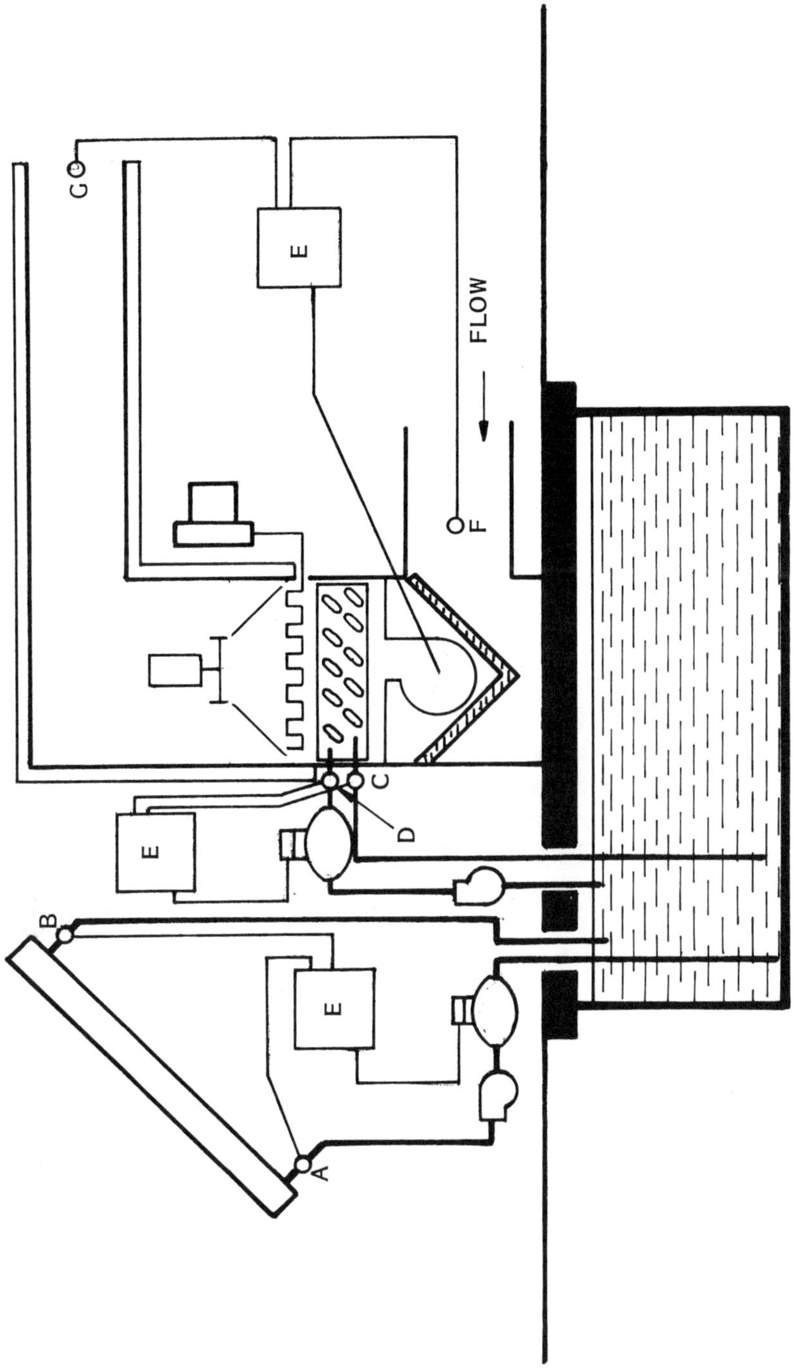

Figure 48. Solar space heating system with three sets of instruments to measure heat output of collector, storage, and total system.

Air Systems

The measurement of energy flow in an air-distribution system requires two quantities: mass flow and bulk temperature rise. Special precautions usually must be taken to ensure accurate measurement of these quantities, such as freedom from extraneous leakage and adequate mixing. Air-duct systems for space heating normally are assembled with little regard for air leakage at the seams, joints, elbows, and transitions. As a consequence, considerable leakage takes place during operation of the system. Normally, this leakage is of little or no consequence to the ability of the air-handling system to perform its normal function. However, when a measure of energy flow is required, the leakage can represent a very significant component of one of the quantities of interest. In most cases, it will be necessary to make certain parts of the duct system leaktight to ensure that the measurements obtained with the instrumentation are representative of the flow. Movable control dampers used to control operating modes are another source of leakage and of possible serious measurement error. Unintentional natural convection and conduction from the storage compartment into the building can also be a significant source of error. The size of these components of error can be kept within tolerable levels if appropriate measures are taken during design and construction of the system.

The presence of a power humidifier, vaporizer, steam injector, or atomizer in a forced-air heating system presents a serious complication to the measurement of heat flow. Since evaporation is an endothermic process, the introduction of water to air stream will generally cause a decrease in temperature, and the energy flow picture would not be complete without a measurement of water evaporated. Certain humidifier models by-pass heated air back to the return-air plenum over a large evaporating surface with the net effect of reducing circulating air throughput and increasing the moisture content with little change in outlet air temperature. All types of humidifiers introduce an energy component that the instrumentation used to record temperature rise and flow rate alone cannot measure. It is preferable to turn off any built-in humidifier and close off any bypass during the measuring period. If water vapor must be provided during the measurement period, it is best to do so by using external humidifiers, such as portable cabinet types employing a vaporizing drum and a fan or room vaporizers that produce steam or a fine mist.

The measurement of temperature rise across the heating device, be it the back-up heater, the solar collector, or the storage compartment, must be representative of the bulk-air temperature rise. If it is possible to achieve a uniform air temperature across the entire return-air duct and across the supply-air duct, then two single-point measurements can be taken to be representative of the bulk temperature rise across the heating plant. To confirm that a uniform temperature has been achieved, it may be necessary to use static and/or dynamic mixers and a long straight duct section to permit uniform laminar flow to become established.

Ideally, the provisions for instrumentation should be specified at the time that the entire system is designed. Such details as the provision for air-tight control dampers where necessary, installation of mixers, provision of straight main and return duct runs without elbows or take-offs over a sufficient length, and air-tight duct sections between temperature sensors should be specified. System design should ensure that energy flows are accounted for under all possible operating modes. In some cases, provisions must be made in the duct layout to eliminate convective air flow, which would otherwise escape detection by the instrumentation. Remember: the instrumentation relies on forced air movement for proper registration.

Measurement of Energy Collected

The indirect method of measuring energy flow described in Chapter 2 may be applied to any branch of a system as long as the air flow is constant during each operating mode.

An electric heating element must be installed in the air loop of interest to perform the calibration; mixing baffles must be installed to ensure adequate air mixing prior to reaching the temperature sensors. Figure 49 provides an example of instrumentation location for a solar collector loop. Most systems automatically switch from one mode of operation to another, such as from charging the storage to drawing from storage during the night. In many cases, the mass flow through some of the air ducts changes because of different path resistance after operation of flow control dampers or auxiliary blowers. When this is the case, the calibration of the energy flow instrumentation must be conducted under each possible mode of operation. Recording information must be kept in separate compartments or channels for each mode, so that the appropriate calibration factor can be applied.

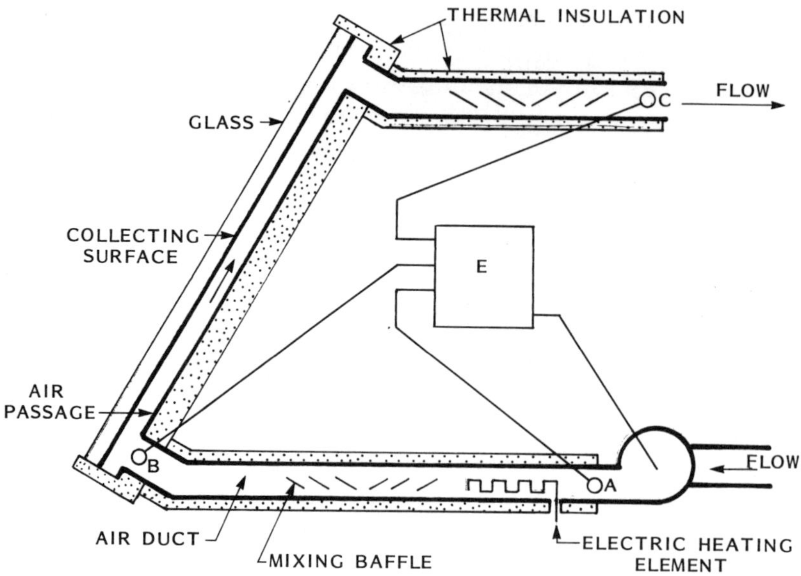

Figure 49. Instrumentation required to measure energy collected by a solar panel in an air-based system.

Three temperature sensors are needed in this arrangement: A and B are used during the calibration to measure the temperature rise of the air produced by the electric heating element; B and C are used during normal operation of the collector to sense the rise in temperature produced by the solar energy absorbed. Mixing baffles must be used to ensure uniform temperature at the temperature sensor locations; this permits the use of single-point temperature measurements to obtain bulk air temperatures. The electronic package E performs the pulse multiplication described in Chapter 2 under "Indirect Method." The calibration usually can be carried out infrequently, unless changes occur in the air resistance due to alterations in settings of dampers or blockage of air filters.

Measurement of Energy Drawn from Storage

Solar systems are usually designed with a provision for storing the excess energy collected during periods when the demand for heat is low. Generally speaking, the amount of stored energy is highest when the need for heat is lowest and vice versa. This being the case, the heat loss from the storage compartment can cause problems of overheating in the building during mild periods, unless special provisions are made for reducing this undesirable loss. Over a heating season, some storage bins may lose a quarter or more of the energy stored via this undesirable mode. The reason for this rather high fraction is not so much the level of insulation around the container but, rather, the long periods of time that the storage may be fully charged and standing-by during mild periods. Predictive controls to limit the level of charge during mild periods usually help to reduce the uncontrolled heat loss to the building.

In installations where the heat loss from the storage compartment enters the heated space, it is advisable to estimate this component of heat gain to the building and credit an appropriate portion to the solar system. However, the portion that causes undesirable *overheating* of the building during periods of no heat demand should not be credited.

Heat flow instrumentation on the heat storage compartment reveals the amount of heat delivered to the building during periods of demand for heat. By comparing this amount of heat with that delivered by the solar collectors over an entire heating season, the stand-by loss component can be calculated by subtraction. Some of this loss will have been useful; some will not have been useful, having caused unwanted overheating.

The instrumentation required to measure the energy drawn from a heat storage compartment is shown in Figure 50 for air systems. The system has three operating modes: delivery of solar energy to the house, charging the storage compartment, and delivery from storage. In the first and last modes, the back-up heater may be called upon to provide additional heat. During retrieval of heat from storage, the control dampers are in Mode 2, which permits air to move through the storage compartment and prevents air circulation to the collector. The instrumentation is similar to that described for solar collectors in the previous section (see Figure 49)—an electric heating element for conducting the calibration, three temperature sensors, air mixing baffles, and thermal insulation. The calibration is valid for a constant air-flow rate. If the rate of air flow is different during various operating modes, separate calibrations and data storage are required for each mode, and data during each mode must be kept separate in order to apply the appropriate calibration factor to each set. Control dampers are normally very leaky and special modification may be required to reduce the leakage. Air leakage in the heat storage compartment, ducting, or control dampers contributes to an error in the energy flow measure-

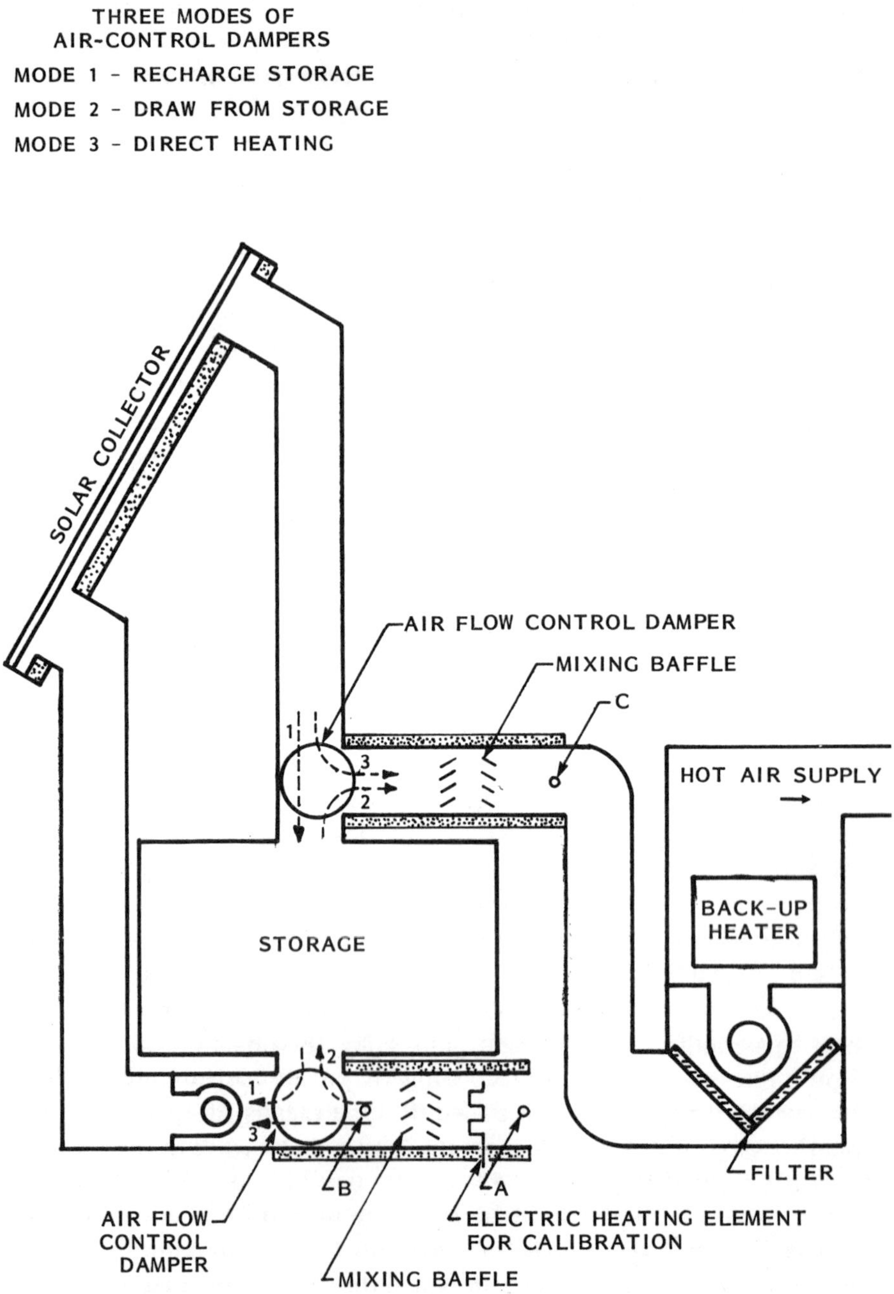

Figure 50. Instrumentation for measuring energy drawn from storage or directly from the collector in an air-based system.

ment. In Mode 3, the dampers cause building air to flow through the collectors and to the building; in Mode 1, they cause air to circulate between the collectors and the storage compartment to recharge the storage material.

LIQUID SYSTEMS

For several reasons, the measurement of energy flows in liquid systems is generally more straightforward than in air systems. Liquid flow meters with the required precision are available and are inexpensive; check and control valves do not normally leak; heat transfer liquid does not escape from the system; and bulk-liquid temperature is easy to obtain by means of thermal wells or immersion temperature transducers.

Measurement of Energy Collected

To measure the amount of energy collected by a liquid-type collector for space heating, a set of instruments can be installed as shown previously in Figure 39 for solar domestic water heaters. The instrumentation consists of a liquid flow meter with a pulsing head, two temperature sensors, and the electronic circuitry to convert the temperature rise to a pulse rate and to gate the pulses according to the flow of liquid, as described under "Direct Method" in Chapter 2. Attention must be paid to convective water flow, particularly during cold sunless periods when heat from storage may drive warm liquid to the collector. This may drive the liquid meter backwards or not affect it (depending on the characteristics of the device). If the meter turns backwards and produces pulses, the electronic circuit will misinterpret this energy flow as a heat gain, since the pulse produced by the flow meter would initiate recording of pulses corresponding to the temperature rise across the collector. One means of eliminating this heat loss and recording problem is to use a check valve in the collector loop to prevent back flow.

It should be noted that the accuracy of measurement in a liquid system does not rely on the rate of fluid flow being constant, as was the case in the air-based system, since the liquid flow meter correctly keeps track of fluid flow by initiating the gating of pulses for recording temperature information. Variation of flow rates during operation in various modes will be taken care of by the flowmeter and the multiplication. The accuracy of the energy-flow measurement will be determined by the same factors discussed in Chapter 2 under "Error Analysis."

Measurement of Energy Drawn from Storage

The instrumentation required for measuring the amount of energy withdrawn from storage consists of components similar to those mentioned above for liquid-type solar collectors. As shown in Figure 51, the temperature sensors are located on the storage tank outlet and on the cold water supply pipe at B and A respectively, and the flow meter is on the return water circuit just ahead of the storage vessel. The electronic circuit E performs the multiplication of temperature rise and flow. The total energy supplied to the house is then the amount drawn from storage plus the back-up heater energy used plus stand-by losses of the tank and piping that were useful. Losses that cause undesirable overheating when space heating is not required should not be credited to the solar system.

SOLAR SPACE COOLING

A typical solar cooling system is shown schematically in Figure 52. It consists of an absorption chiller, which is supplied with hot water by a solar collector system with some storage capacity and a back-up heater for operating during sunless periods. It is supplied with cooling water from a cooling tower. The chilled water circulates through a heat exchanger in the building's air distribution system to cool the air. To measure the amount of heat being removed by the chilled water, the energy flow instrumentation described in Chapter 2 under "Liquid Systems" can be used and installed as shown in Figure 52. The instrumentation consists of two temperature sensors located at A and B, a liquid-flow meter with pulse output capability, and the electronic package to convert temperature difference into a pulse rate and to multiply by liquid flow to compute energy flow.

Measuring energy flow is much simpler on the water side of the

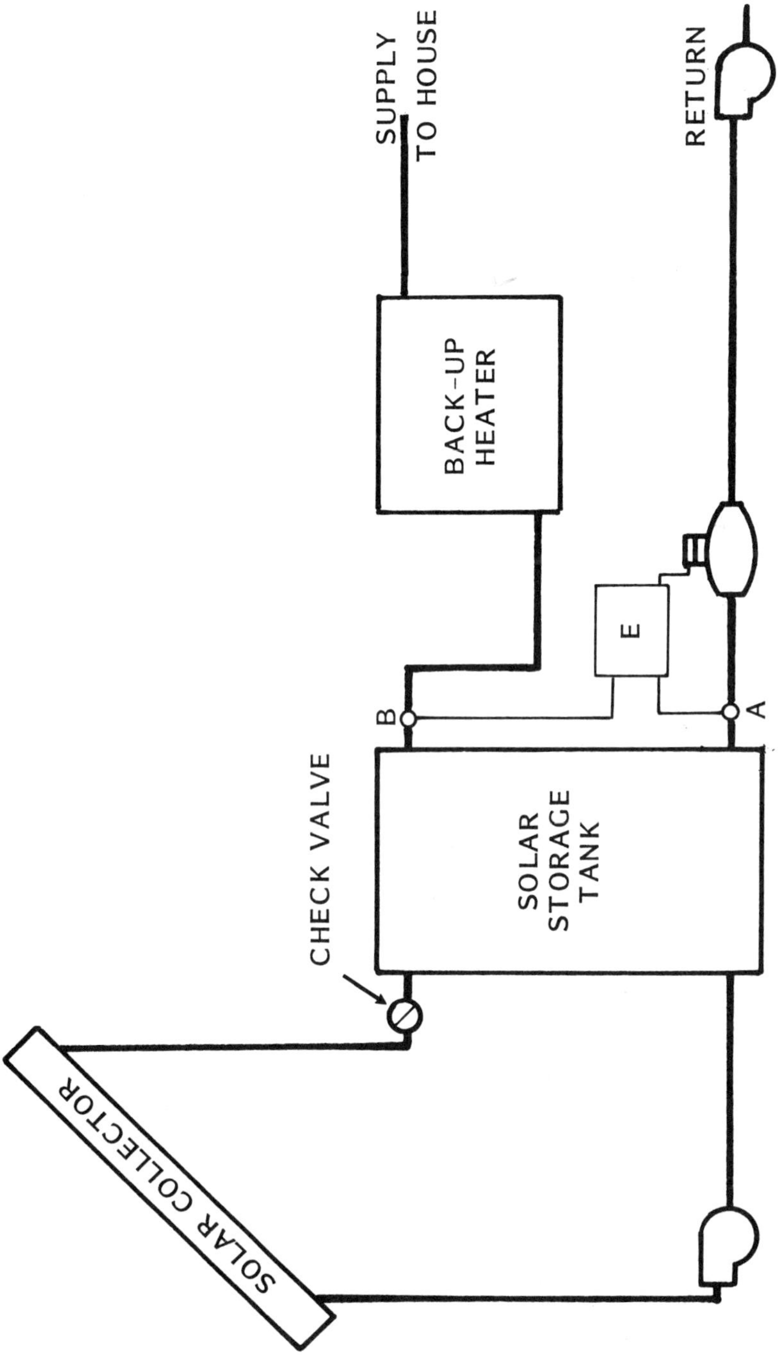

Figure 51. Instrumentation for measuring heat drawn from storage in a hydronic space heating system.

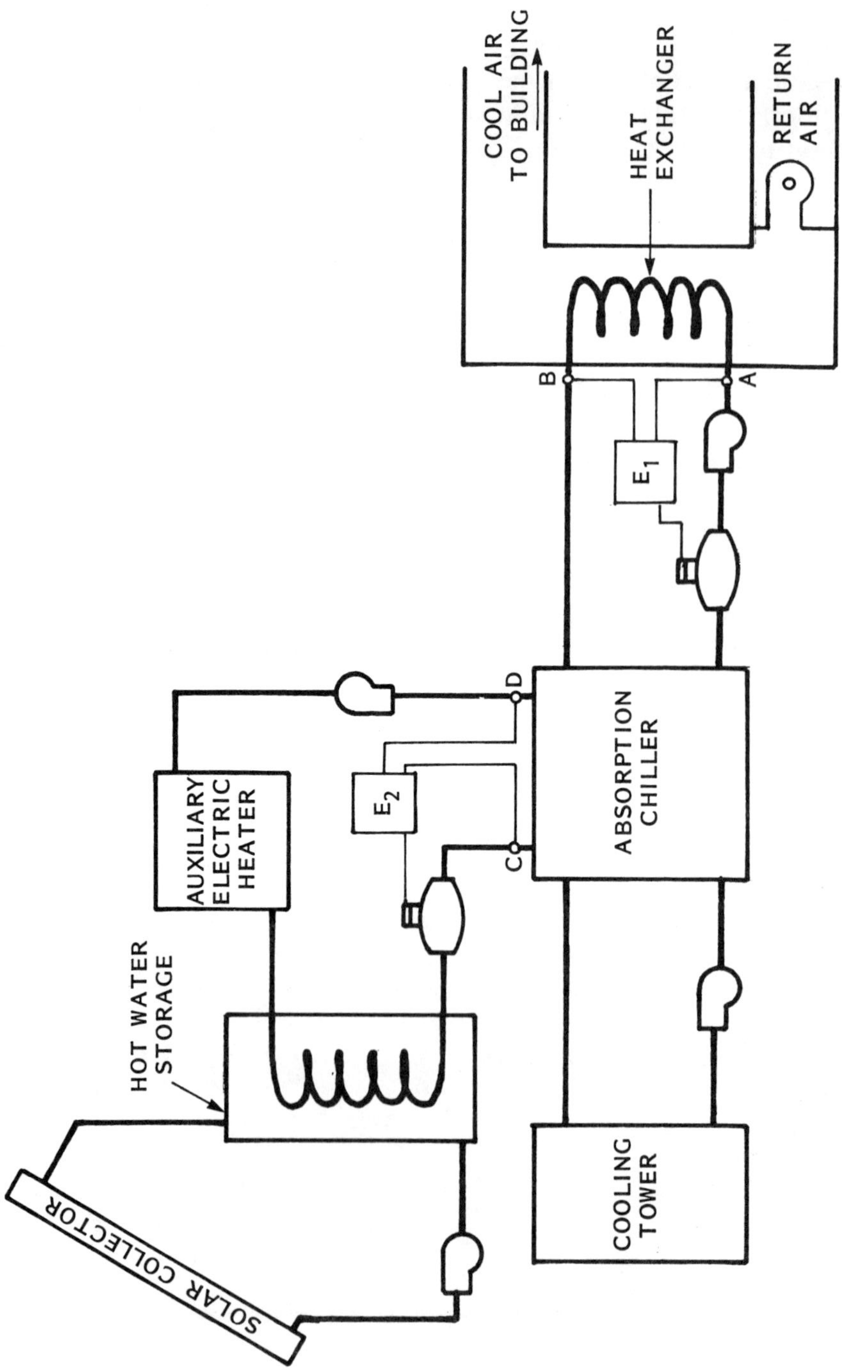

Figure 52. Instrumentation to measure energy flow in a solar cooling system using a lithium-bromide water absorption chiller.

heat exchanger than on the air side because the air may change in moisture content during cooling, requiring the measurement of enthalpy before and after cooling. The measurement of energy flow in the chilled water loop is useful to determine cooling capacity at various operating temperatures of the collector, cooling tower, and building air or at various rates of fluid flow. To establish the coefficient of performance (COP) of the cooling unit, the heat energy input from the solar heating system must be quantified as well as the amount of energy absorbed by the chilled water:

$$\text{COP} = \frac{\text{Energy absorbed by the chilled water}}{\text{Energy input from the solar heating system}}$$

The energy input can be measured by installing instrumentation on the solar loop similar to that specified for the chilled water loop, namely two temperature sensors at C and D, as shown in Figure 52, and a flow meter in the loop. A second electronic package will produce pulses representative of energy input. By totalizing pulses over, say, 15-minute periods, the average COP over that period can be quantified. Similarly, by totalizing pulses over a day, the daily average COP can be obtained.

4

The MIMIC Principle

INTRODUCTION

The output capability and efficiency of heating systems that make use of renewable energy—solar water heaters, heat-pump water heaters, and space-heating heat pumps—depend, to a certain extent, on the amount of energy required of the system, on when the energy is required, and on outdoor conditions such as temperature and amount of sunshine. For example, consider two identical solar-assisted water heaters operating in two houses side by side on a sunny, warm day. Suppose that during the entire day one house uses 91 liters (20 gal.) of hot water and the other 450 liters (100 gal.). The performance of each could be as shown in Table 6.

TABLE 6
Performance of Two Identical Solar-Assisted Water Heaters

	Volume of Water Used (liters)	Equiv. Energy Content of Water (kWh)	Electricity Used (kWh)	Solar Energy Used (kWh)	Percent Solar Contribution (percent)
House A	91	6	½	5½	95
House B	450	30	20	10	33

These hypothetical cases illustrate that when little hot water is used, as is the case in house A, the percentage contribution of the solar system is very high, but the amount of energy delivered may be small. Once the capacity of the storage is filled, solar energy collected is lost to the environment. On the other hand, when a large quantity of hot water is used, the solar system delivers more energy, but in terms of percentage contribution to the total requirement of the house, it may be small, as illustrated in the case of house B. For this reason, systems whose performance or efficiency is affected by use pattern and by weather must be tested under a variety of operating and use conditions to develop reliable comparison and application data. This requirement normally calls for testing over a considerable length of time. Consequently, to minimize the cost of the study, it is highly desirable to conduct the tests on a system that requires a minimum amount of attention. Complete water- or space-heating (cooling) systems can be installed and tested in the laboratory under simulated operating conditions and can be instrumented readily to provide performance data as needed. To obtain a realistic measure of performance, however, the system of interest must be instrumented and observed over a period of several months while operating in a real fashion. This is particularly true for solar water heaters and heat pumps, whose performance characteristics vary widely with the season and with demand for energy.

Besides the expense of installing experimental heating systems in residences, the complications of installing instrumentation and collecting records, providing a 24-hour maintenance service, and, possibly, having problems in relations with the residents due to breakdowns in experimental equipment, it is usually necessary to repeat the test in several locations to obtain a better picture of the effects of usage on performance. Proper field testing is very costly and time-consuming.

One would be able to test heating equipment at lowest cost if it were possible to conduct the "field test" in the laboratory, duplicating the use pattern in real time during the same weather conditions as in the house. One such method can be applied to a number of different types of space- and water-heating systems. In addition to quantifying the solar contribution to the total energy delivered, various other useful measurements can be readily obtained. The method involves installing the test equipment in the laboratory and telemetering real use information from an actual sample house equipped with a "normal" heating or cooling appliance such as a conventional electric water heater or a central electric furnace for space heating.

For domestic water heaters, the instrumentation at the laboratory is designed to control the operation of the test heater at the laboratory so as to duplicate the operation of a conventional water heater at the house, or to MIMIC the operation. A telephone link between the house and laboratory is required, along with simple meters with pulse outputs to transmit electricity-use and water-use data to the laboratory. At the laboratory, a receiving circuit controls the draws of water from the experimental water heater (e.g., solar-assisted or heat-pump water heater) to duplicate the usage at the house. The amount of electricity used by the water heater in the house can then be compared with the smaller quantity used by the solar-assisted water heater in the laboratory.

To use the MIMIC technique for testing experimental space-heating equipment, such as heat pumps or solar systems, in the laboratory it is necessary to transmit energy-use data from the "test house" equipped with a central electric furnace, in a manner similar to that described for water heaters. The pulses transmitted by the meter on the electric furnace are received at the laboratory, where the operation of the experimental equipment is controlled to duplicate the heating demand at the house. This technique has the advantages of low cost for field testing, flexibility, and little inconvenience to homeowners. The solar system under test can be installed at a convenient test site or laboratory and its simulated performance at various houses quantified by simply arranging to transmit the use data from these houses to the solar system in the laboratory. The configuration required is shown on Figure 53.

The instrumentation required on a house furnace to produce the signal required for a MIMIC operation is minimal: a conventional kWh meter with pulse output and an indicator for the status of the space-heating thermostat. The two signals are imposed on a telephone pair (solid copper pair of telephone wires is adequate for several miles of transmission) with a ground return. To produce the signals required to MIMIC the operation of a water-heater, it is necessary to install a kWh meter with pulse output on the electric supply and a water meter (billing type) also with a pulse output.

At the laboratory or test site, the telephone signals operate two receiving relays. A microcomputer or controller produces a control signal, while it monitors the energy delivered by the experimental system. As soon as the energy delivered by the laboratory appliance matches that used by the remote appliance, a control signal is sent to the laboratory appliance to turn off. Because hot water for domestic use tends to be drawn in an unpredictable manner, and often in many small

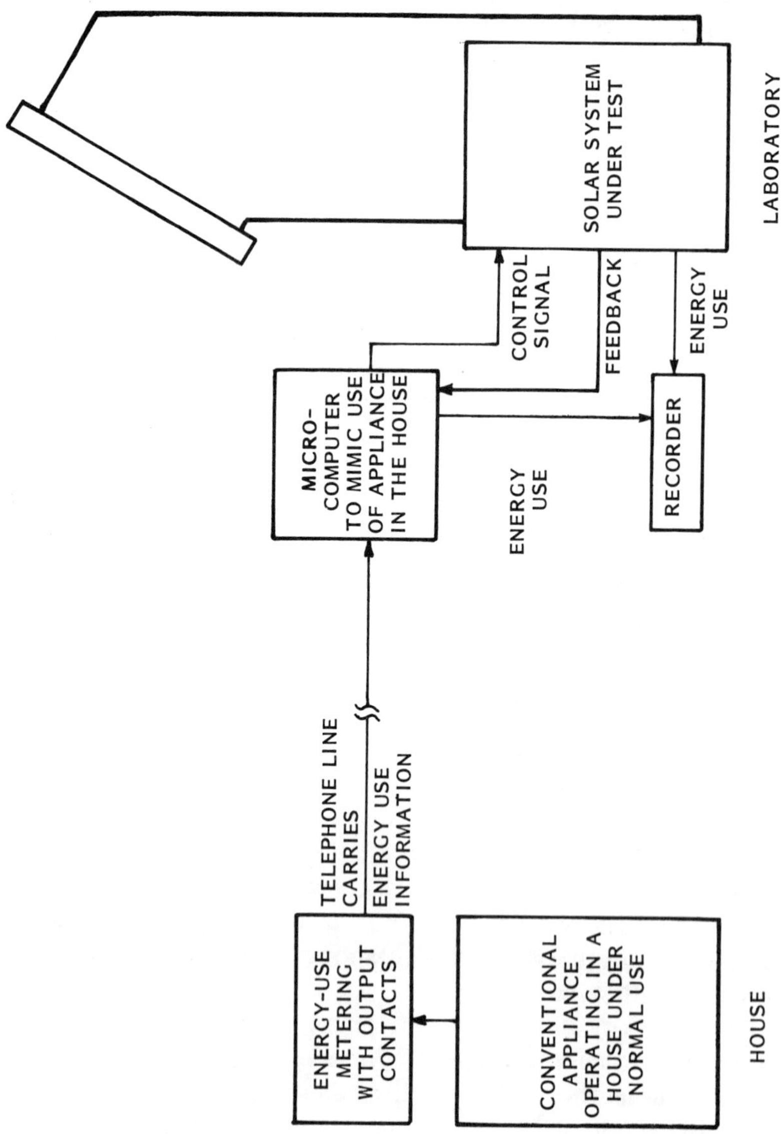

Figure 53. Configuration required to MIMIC the operation of an appliance by a solar system under test at a remote location.

quantities, the remote data can be collected for 15 minutes, and then used to trigger the laboratory system once each 15-minute period to dump an equivalent volume of hot water. This deviation in use has a very small effect on system performance, but it must be kept in mind when drawing conclusions on the electrical demand of the MIMIC system as a function of time.

SOLAR WATER HEATING

The instrumentation required to operate a water heater in a MIMIC mode will be described in detail in this section. The solar water-heating system under study is installed in the laboratory as a preheater to a conventional electric water heater. A household with an electric water heater is selected as the "test site" and is instrumented as indicated in Figure 54. A water flowmeter with pulse-output capability is installed on the cold water supply line to the tank. The meter is to produce one pulse per given volume of water drawn, such as ¼ liter. A kilowatt-hour meter is installed on the power supply to the water heater. This meter is also to have a pulse-output capability (one pulse per revolution of the disc). The pulses are to be transmitted to the laboratory via a single telephone pair to be picked up by two receiving relays and monitored by a dedicated circuit or microcomputer at the laboratory. Similar meters with pulse output are installed on the cold-water supply line and on the electrical supply circuit to the experimental water heater in the laboratory. An integrating solarimeter with pulse output, as described in Chapter 1, is also installed at the laboratory to record solar incident radiation. Hot water from the experimental system is drawn by means of operating a solenoid valve at the outlet of the water heater. The microcomputer is programmed to accumulate data over a 15-minute period and then to open the solenoid valve to dump a volume of hot water in the laboratory equivalent to that used at the house. The cycle is repeated every 15 minutes. The 15-minute (maximum) delay between the draws at the house and the dump at the laboratory will have a minimal effect on the performance of the solar system. The microcomputer also can be programmed to accumulate pulse data from the meters for 15-minute periods and store the readings for daily printing or display. At least four variables are recorded at the laboratory: electricity use at the house, electricity use at the laboratory, hot water use,

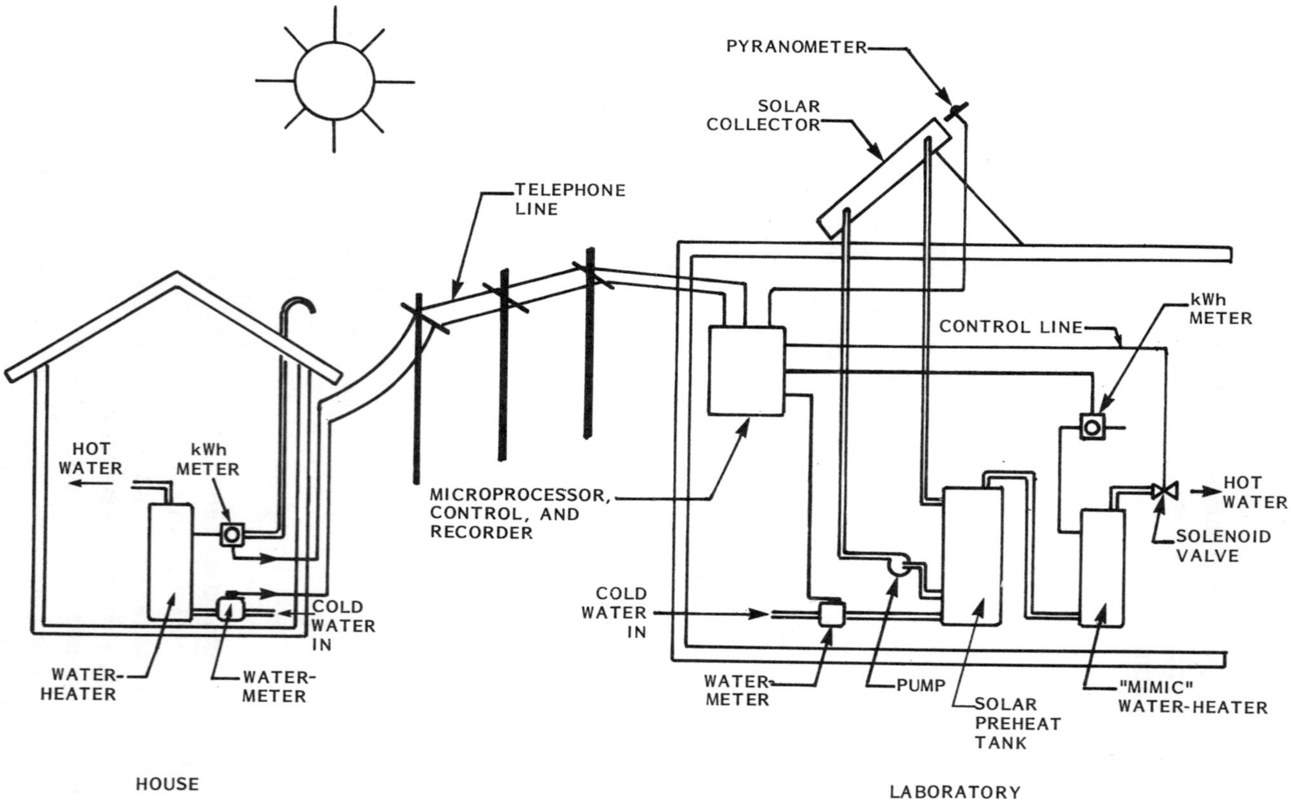

Figure 54. Instrumentation required to MIMIC the operation of a water heater. Reprinted with permission from the American Society of Heating, Refrigerating and Air Conditioning Engineers, Atlanta, GA.

and the solar energy available. As a check on the flow-control system, it is also desirable to record hot water drawn at the laboratory, to make sure it matches exactly what was drawn at the house. In this manner, it is not necessary to install and service recorders at the house; all the necessary data from the house is available at the laboratory.

A block diagram of the main components of the MIMIC control and monitoring system is shown in Figure 55. The blocks on the left side of the diagram correspond to the electric and water meters installed on the conventional water heater at the house. The output contacts operate two relays, which transmit pulses to the laboratory via a single telephone pair. The signals are picked up by two receiving relays at the laboratory. The energy data from the house and from the device under test at the laboratory are recorded directly by an appropriate recorder, such as a magnetic tape "demand recorder" or an electromechanical totalizing or printing counter. The water-use data from the house and the laboratory are fed into the microcomputer. This device is programmed as a comparator. It totalizes the number of water-flow pulses arriving from the house during 15-minute periods and then opens the solenoid, which dumps water from the system under test. As soon as the number of pulses from the meter on the test system matches the draw of the remote water heater, the microcomputer turns off the solenoid and waits another 15 minutes. In this manner, the system under test in the laboratory MIMICS the operation of the water heater in the house. For more information on the program, the operation of the microcomputer, and the operation of the control circuit see Appendix B.

The solar contribution to the test system at the laboratory shows up as a reduction in electrical energy used when compared to the energy consumed at the house. Before the comparison can be made, however, the thermostats on both electric water heaters must be adjusted to produce the same outlet water temperature, so as to match exactly the amount of energy used by both water heaters when the solar assist is bypassed.

The consumption data can be totalized daily and plotted versus time. Sample data from a MIMIC system operating at a laboratory imitating the hot water usage of a water-heater 7 km away are shown in Figures 56 and 57 for fall and winter months, September to December and January to April, respectively.[11] Presented in the same figures are the daily water usage, electrical consumption of each system, solar contribution, and daily total incident radiation on the solar collectors. As

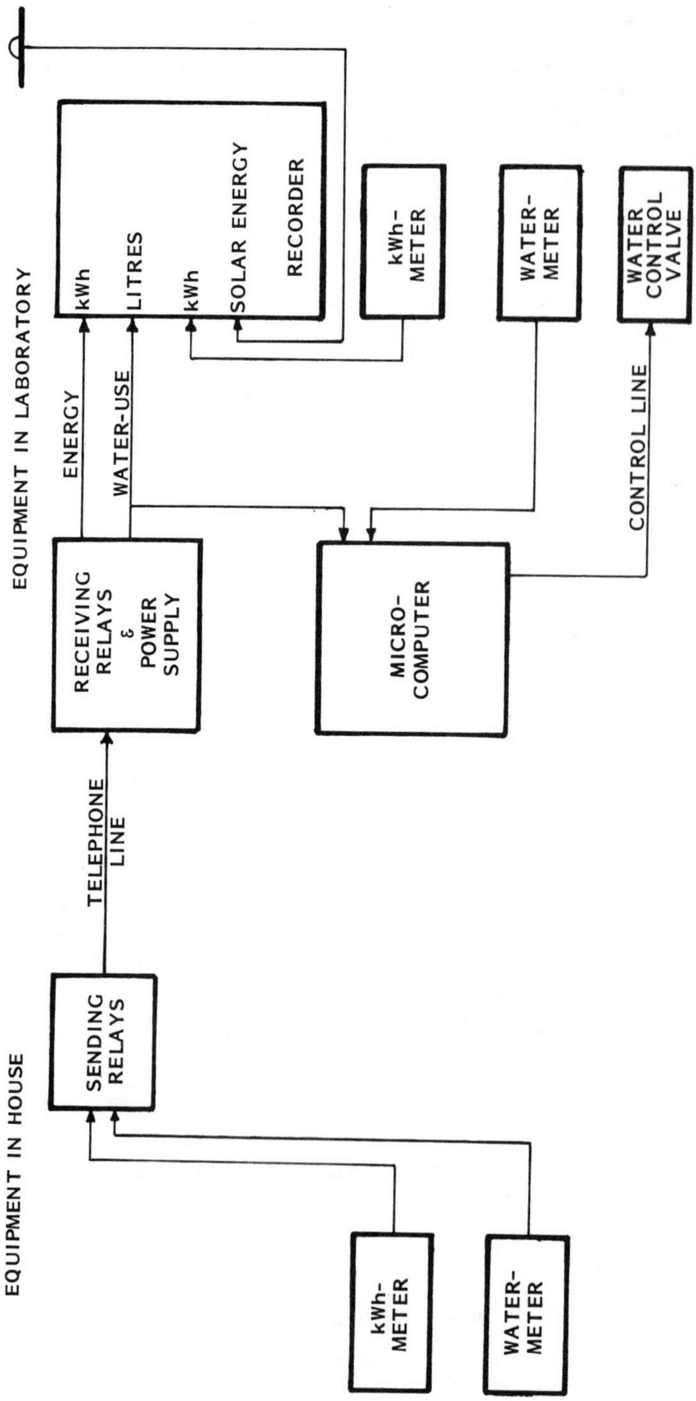

Figure 55. Block diagram of the MIMIC control and monitoring system.

can be seen from the electrical consumption records, the period to September 26th was used for matching the settings of thermostats with the solar assist bypassed. From the 27th on, considerably less energy was used in the laboratory because of the contribution from the solar storage tank, which was well recovered by that time. The solar contribution is seen to decline towards December because of the lower solar energy availability. It is seen to increase again toward Spring.

Electrical demand can readily be recorded by means of a multichannel magnetic demand recorder, since the electrical-use data are already in the form of contact closures. These recorders are often used by electrical utilities to collect records for demand billing or for load surveys. The magnetic tape records are usually translated into a sequence of average demands over fixed time intervals (such as 15 minutes or one hour). See "Recording Solarimeter Output" in Chapter 1.

The difference in the demands for electricity observed between the two water heaters provides a direct indication of the actual impact of solar energy on electricity demand. The comparison may be carried out on a 15-minute, hourly, daily, or any other basis.

Examples of 15-minute demand data obtained from the installation described above are presented in Figures 58 and 59. Two days have been selected, one with a high solar contribution and one with a low solar contribution. It is interesting to note that the demand for electricity occurs at about the same time in both the solar-assisted and the conventional water heaters, but the duration is less for the solar-assisted unit, especially during the day of high solar input. Electric power utilities interested in obtaining information on the impact of solar assist on the demand for electricity and energy use would benefit from the direct comparison that the MIMIC system offers. To obtain reliable diversified demand information, a considerable number of such twin installations would be required—possibly about 20.

The evaluation of performance of heat-pump water heaters in the field can also be conducted in the laboratory by means of the MIMIC principle and instrumentation and controls similar to those described above for solar water heaters. Although heat pumps have been around for over 50 years, it is only recently that they have been applied to heating domestic water. The attraction is the lower cost of energy brought about by a usual efficiency of 200 to 400 percent. The amount of energy required for heating water in a household is second only to that required for space heating in many parts of the continent. Consequently, it is receiving a considerable amount of attention in this period of increasing fuel prices.

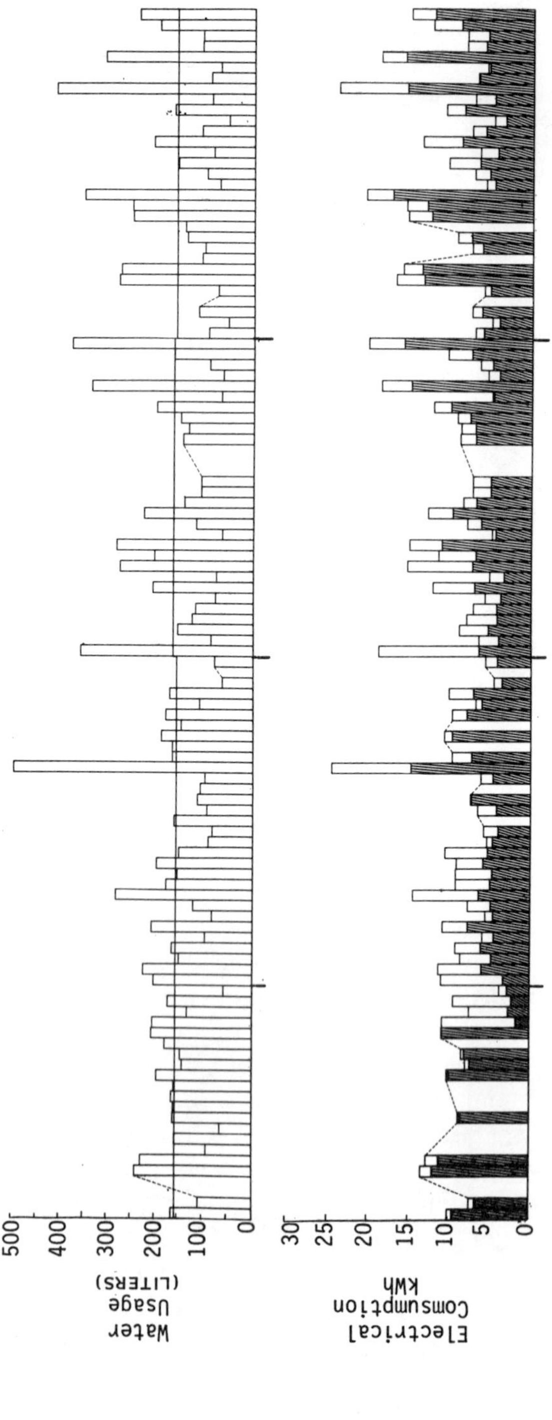

(continued)

Figure 56. Hot water use, electrical consumption of conventional and solar-assisted water heaters, and incident solar energy—September–December, 1978. Reprinted with permission from the American Society of Heating, Refrigerating and Air Conditioning Engineers, Atlanta, GA.

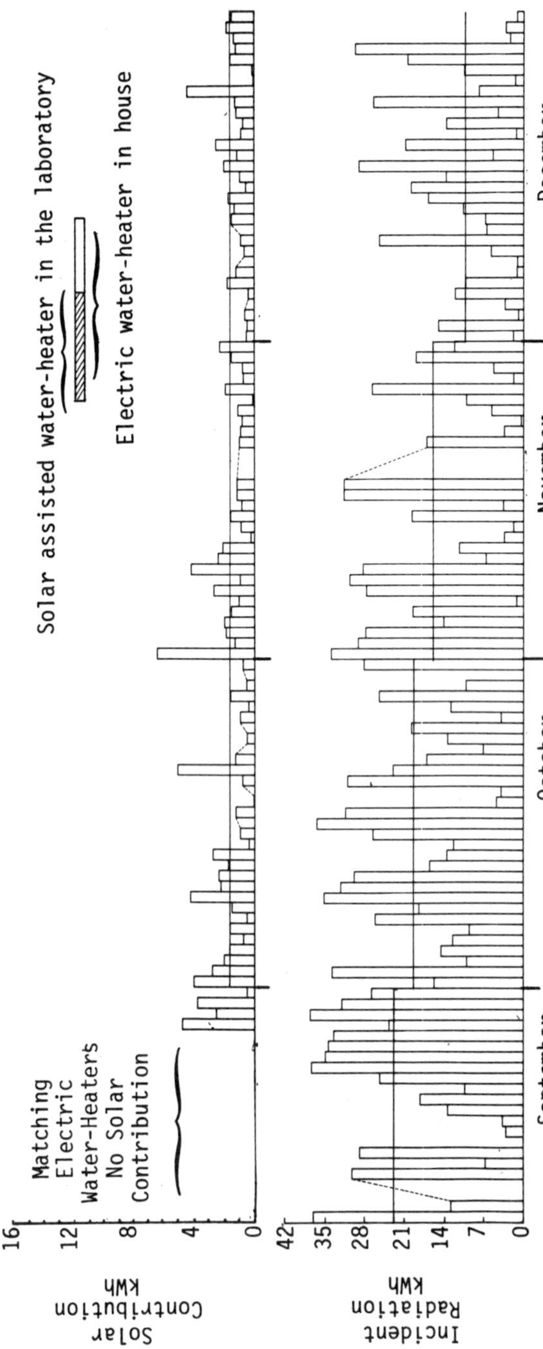

Figure 56 (continued)

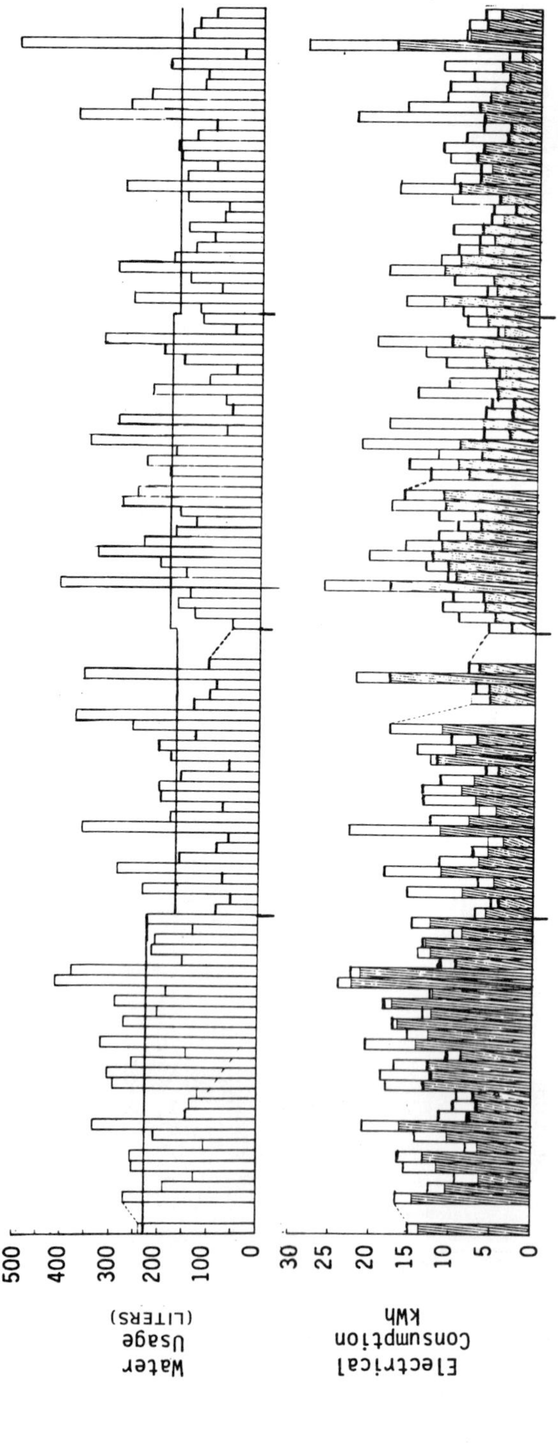

(continued)

Figure 57. Hot water use, electrical consumption of conventional and solar-assisted water heaters and incident solar energy—January–April, 1978. Reprinted with permission from the American Society of Heating, Refrigerating and Air Conditioning Engineers, Atlanta, GA.

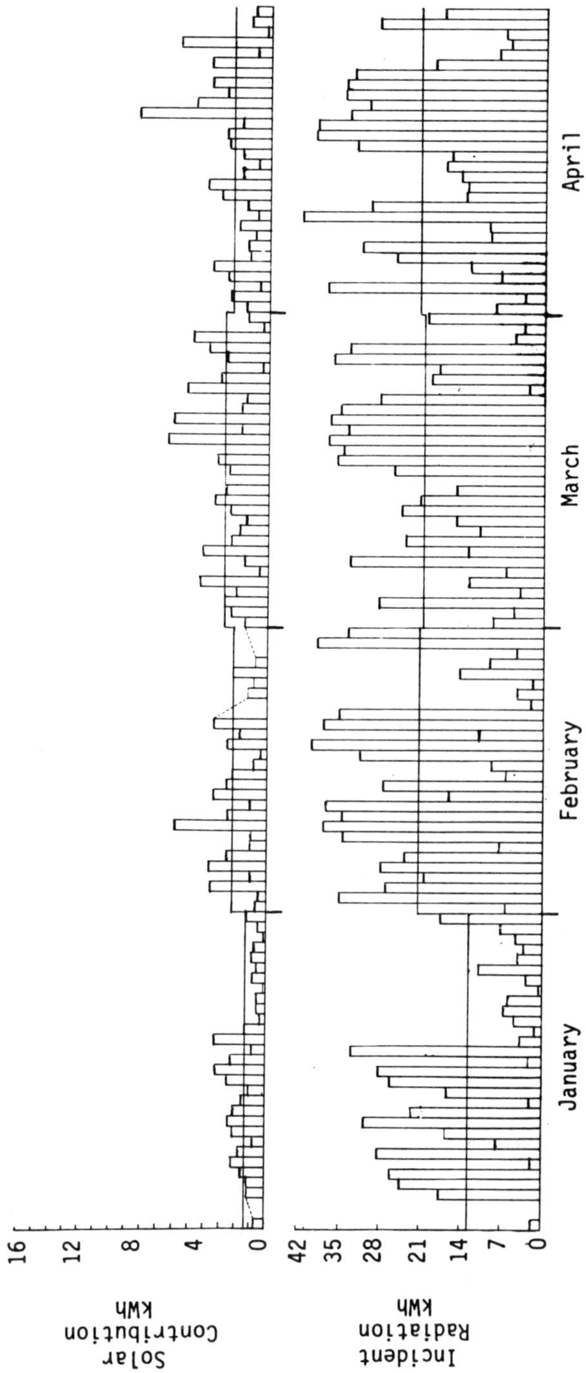

Figure 57 (continued)

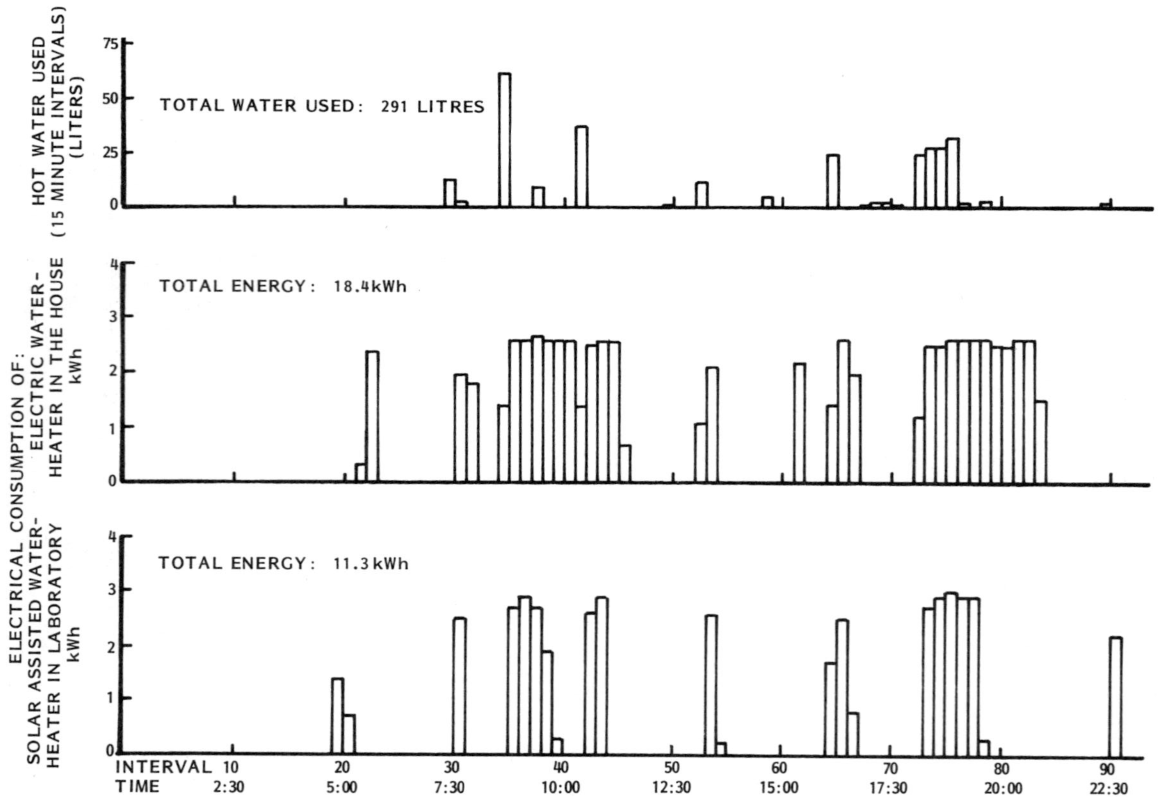

Figure 58. Hot water demand and electricity used by water heaters at the house and at the laboratory over one day, 15-minute periods—sunny day. Reprinted with permission from the American Society of Heating, Refrigerating and Air Conditioning Engineers, Atlanta, GA.

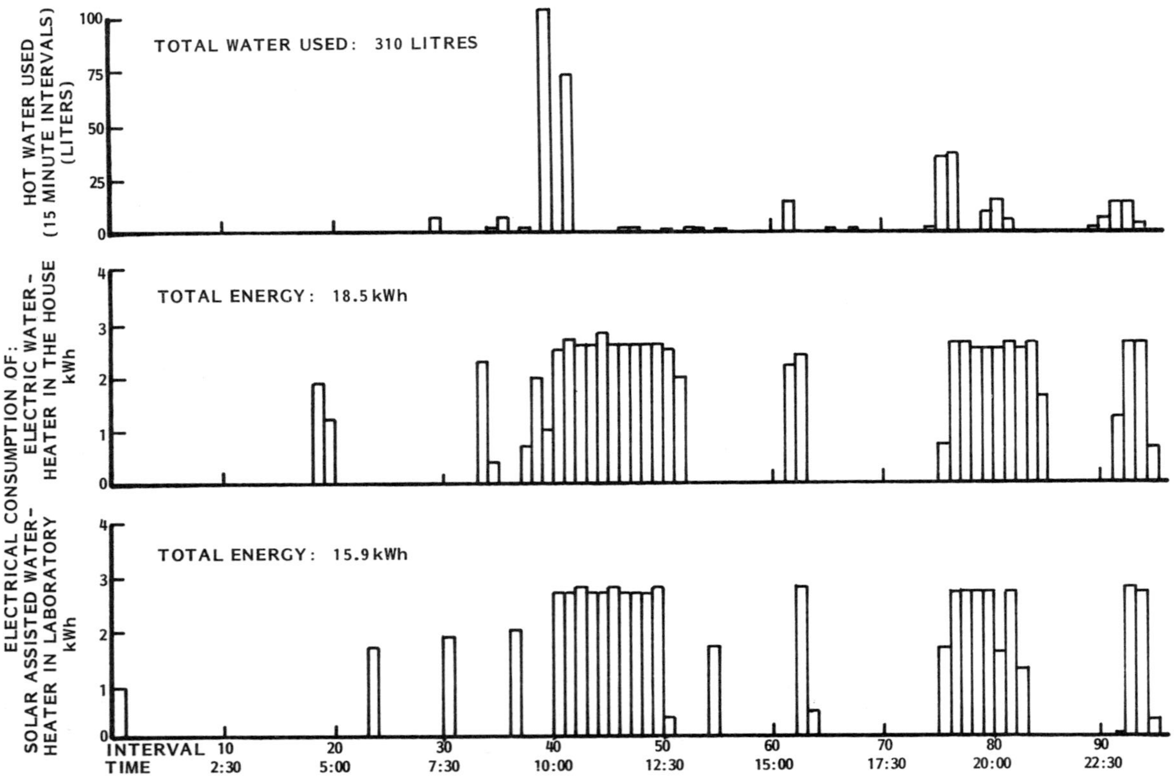

Figure 59. Hot water demand and electricity used by water heaters at the house and laboratory over one day, 15-minute periods—cloudy day. Reprinted with permission from the American Society of Heating, Refrigerating and Air Conditioning Engineers, Atlanta, GA.

Heat-pump water heaters are taking many forms. Some use indoor air as a source of heat, others use outdoor air. Some systems recover heat from household waste water, others extract heat from well water. Four different configurations are illustrated in Figure 60. A backup tank with conventional electric-resistance heating can be installed downstream of the heat-pump water heater to ensure that water is delivered at the desired temperature. Some heat-pump water heaters have a backup heater near the top of the preheat tank and rely on stratification of hot water at the top of the tank to prevent mixing. The heat pump provides heat to the lower portion of the tank by means of an outlet with a diffuser just below the heating element and an inlet near the bottom, as illustrated in Figure 61. Some heat pumps are now capable of delivering water at close to 60°C without the aid of an electric-resistance heating element. The last two types of heat-pump water heaters are not suitable for MIMIC operation because the outlet temperature is not likely to remain virtually constant during use. This was a primary requirement for "matching" use of electrical energy while the assist is bypassed.

The efficiency of a heat pump depends on various factors, including the temperature difference between the hot water at the outlet of the heat exchanger and the air or water temperature at the source. The efficiency of a heat pump will depend not only on the temperature of the air or water source but also on the thermostat setting and the use of hot water.

The MIMIC method can be used to direct the operation of a laboratory heat-pump water heater to MIMIC the operation of a remote conventional electric water heater in a house. The instrumentation and controls would be nearly identical to those required for solar water heaters.

In each case, the electric backup water heater must be of the same capacity and wattage and set to the same temperature as the water heater at the house. Furthermore, the electric tank in the house must never run out of hot water, and the preheater tank in the laboratory must never heat the water above the temperature of the backup tank, or else errors in the interpretation of water and energy flows will occur.

Since the MIMIC arrangement requires that the same amount of energy be used for a given amount of water usage (with the solar system bypassed), the standby loss of each system is included in the measured energy. It should be noted that the standby loss of the solar tank (in a two-tank system) is made up by the sun. Also, although the level of insulation in the electric tanks in the house and in the laboratory

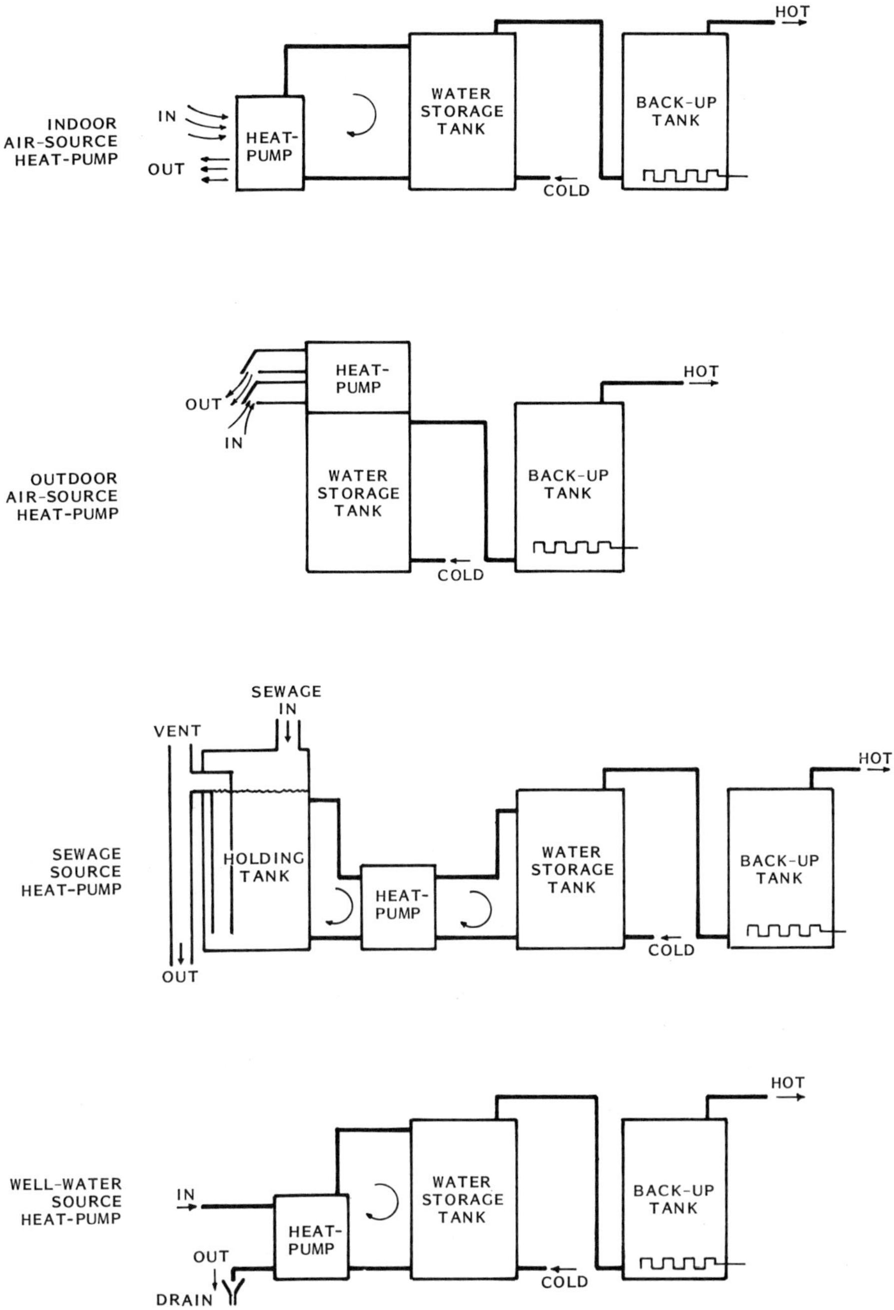

Figure 60. Various configurations of heat-pump water heaters.

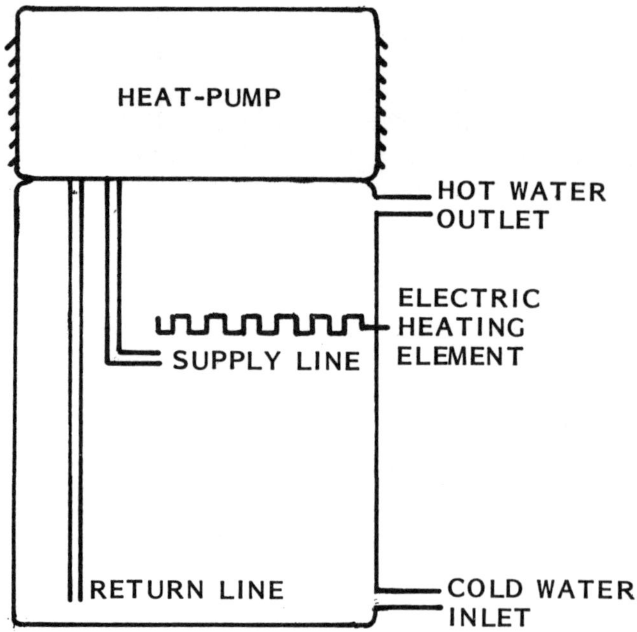

Figure 61. Packaged heat-pump water heater.

may be identical, the heat loss of the tank in the house may tend to be lower than that of the laboratory tank because of the length of time that the house tank may be partly recovered. Since the recovery period of a solar-assisted tank is shortened by the preheated supply of water, this tank would be fully recovered over a longer period than the unassisted tank in the house. If the water usage is known, the reduction in heat loss may be calculated. Well insulated tanks of 180-liter capacity have a heat loss of less than 100 watts. The difference in standby loss has a very small effect on the assessment of performance.

HEAT PUMPS FOR SPACE HEATING

The most common heat pump for space heating in North America is the air-to-air type, which uses outdoor air as the heat source and delivers the heat to a central forced-air distribution system. In fact, the original heat source is the sun, and one could consider the heat pump to be another type of solar-assisted heating system. The "collector" is outdoor air, which, after absorbing heat from the sun, is drawn over the outdoor heat exchanger, which in turn picks up some of the heat energy. The heat-output capability of the heat pump declines as the outdoor temperature drops, and, in normal applications, a secondary heat source is required as backup heating during cold weather. The secondary source is usually electric-resistance heaters or gas-fired or oil-fired furnaces. Heat-pump installations are often sized according to the air conditioning capacity required for summer cooling. In areas with mild winters and hot summers, the heat pump can provide all of the heating energy required. In areas with cold winters, the heating capability of the heat pump will be adequate down to a temperature called the balance point temperature, below which the heat pump alone cannot maintain the required house temperature. The backup heating system then must be put to use. Electric heaters installed downstream of the heat exchanger in the furnace can be energized at temperatures below the balance-point temperature to supplement the heat being produced by the continuously operating heat pump. It is desirable to keep the heat pump operating below the balance point temperature because its efficiency is still high, and the heat it delivers costs less than that delivered by electric-resistance heating.

Oil-fired or gas-fired furnaces used as backup heating to a heat pump must be operated in a different manner. In these installations, the heat exchanger of the heat pump is installed downstream of the combustion chamber, and simultaneous operation of both the heat pump and the burner cannot be permitted because the heat pump is not capable of delivering heat to a hot-air stream. This problem could be eliminated if it were practical to install the heat exchanger of the heat pump upstream of the combustion chamber. This practice is not permitted, however, because of condensation that can take place in the combustion chamber during summer air conditioning. The chilled air flowing around the combustion chamber in the summer can cause condensation inside the fire box due to humid outdoor air finding its way down the chimney. Condensation can lead to corrosion and early failure of the combustion chamber. Operation of the heat pump below the balance point temperature with a fuel-fired backup furnace is therefore not practical. The contribution of the heat pump to the heating requirements of the building while operating in this limited mode is smaller than could be the case with electric backup heating installed as usual downstream of the heat exchanger.

The performance of a heat pump is very much dependent on the heating demand imposed on it, on the backup system and controls used, on the balance point temperature of the installation, on outdoor conditions, and on other factors. To measure the performance, energy output data must be collected over long periods of time, as described in Chapter 2, Indirect Method, Air Systems. Since the instrumentation required for measuring heat output is rather involved and the performance of a specific heat pump may be different from one installation to the next, it may be more convenient and less costly to install and instrument the heat pump in the laboratory and operate it so as to match the output, or MIMIC, the operation of a heating system in a nearby house. The operation of the heat pump can be controlled by signals from a house heated by a central forced-air electric-resistance furnace. The heat pump may be "moved" to another house merely by obtaining control signals from the central electric furnace in another house.

The operation of the furnace in the house can be telecommunicated to the laboratory by the arrangement shown in Figure 62. A kWh meter with a pulse-output capability is installed on the supply to the furnace and its signal is transmitted by a dedicated telephone pair to the laboratory, along with an indication of thermostat status (open or closed).

At the laboratory, the heat pump is installed in the normal manner (with outdoor unit outdoors and indoor unit indoors) and supplied with return air at approximately the same temperature as the air supplied to the furnace in the house. The outlet air is dumped into the laboratory after its energy content is measured. The instruments and controls in the laboratory are arranged to operate the heat pump so as to match the energy used by the electric furnace at the house. If the house is not too distant from the laboratory, the outdoor section of the heat pump and the house will be exposed to the same weather, and the heat pump will behave as if it were installed in the house. Matching the heat output of the heat pump to that of the furnace requires that the heat output of the heat pump be measured continuously. This can be done most conveniently by the indirect method for air systems described in Chapter 2 and illustrated in Figure 24.

The heat pump at the laboratory is controlled by a microcomputer operating as a comparator circuit (see the flow diagram in Figure 63).

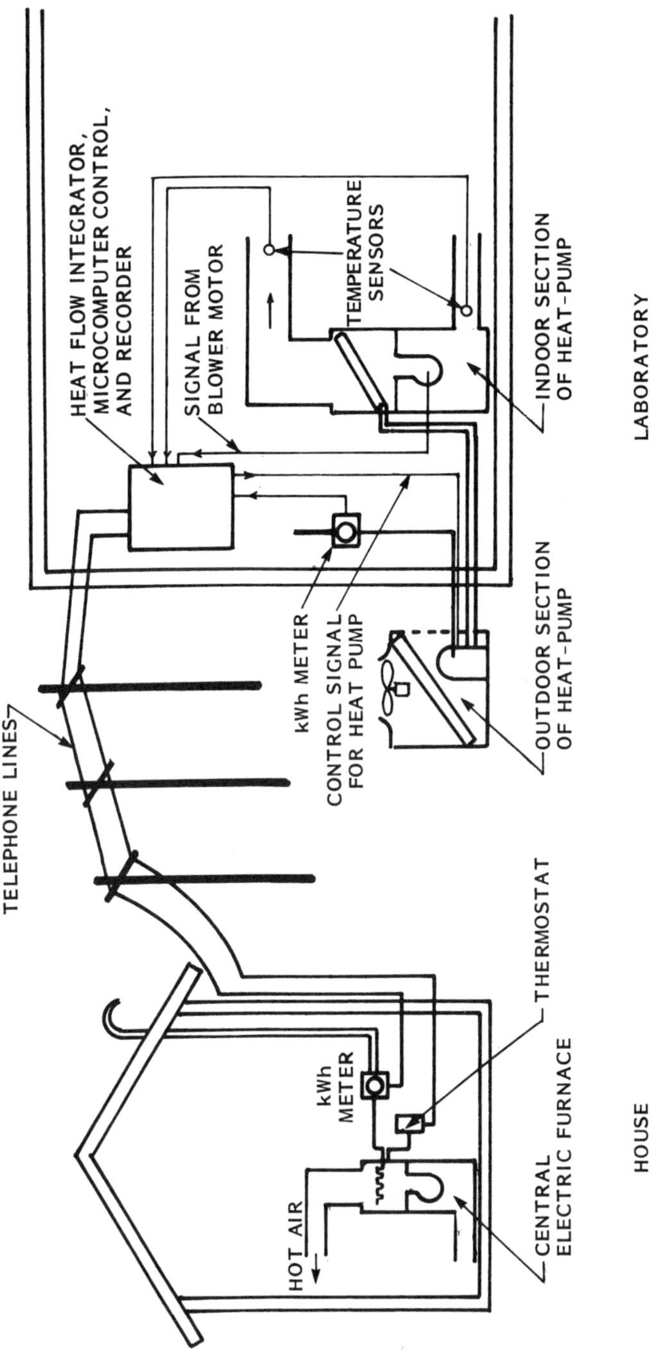

Figure 62. Instrumentation required to MIMIC the operation of a space-heating furnace.

Via the telephone link, this circuit receives pulses from the kWh meter on the electric furnace supply at the remote location and the status of thermostat contacts (open or closed). As soon as the thermostat at the house closes and pulses from the kWh meter begin to arrive at the laboratory, a control energizes the heat pump. Pulses arriving from the house are *added* to the register. While the heat pump operates, its measuring circuit produces pulses in proportion to the heat energy being delivered. These pulses are *subtracted* from the register until the register reads zero again, at which time the heat pump is stopped. The calibration of the pulse-rate converter for the heat pump is arranged so that each pulse from the meter in the house and from the instrumentation on the heat pump in the laboratory represents the same amount of energy. In this manner, the heat pump "matches" the energy output of the electric furnace at the house, or MIMICS the furnace, then waits for the next furnace cycle. In general, during mild weather the heat pump at the laboratory will be able to keep up with the furnace at the house. This mode is equivalent to operation of the heat pump above the balance-point temperature.

During cold weather, the heat pump may be unable to match the energy delivered by the furnace, in which case the heat pump will operate below the balance-point temperature. The thermostat in the house may call for heat again while the heat pump at the laboratory may not yet have matched the amount of energy indicated by the previous furnace cycle. In that case, the balance "owing" by the heat pump must be tallied in a separate register. Refer again to Figure 63, left-hand branch. This amount of energy corresponds precisely to the amount of back-up heat required by the heat pump had it been operating in the house. A recorder or a pre-programmed microcomputer can be used to keep track of daily or periodic consumption figures for the furnace, heat pump, and the tallied "back-up" energy required.

Air-source heat pumps operating at outdoor temperatures between +5°C and −5°C accumulate frost on the surface of the outdoor coil. As the frost builds up, air flow through the coil drops, and the performance of the unit declines until a defrost cycle is initiated automatically by a control or timer.

The defrost operation is performed most commonly by switching the system to the summer cooling mode; that is, by reversing the refrigerant cycle to cool the indoor air and heat the outdoor coil for a short time. During this period, which may last about 5 minutes and be repeated as frequently as every 5 hours or so, the heat pump draws

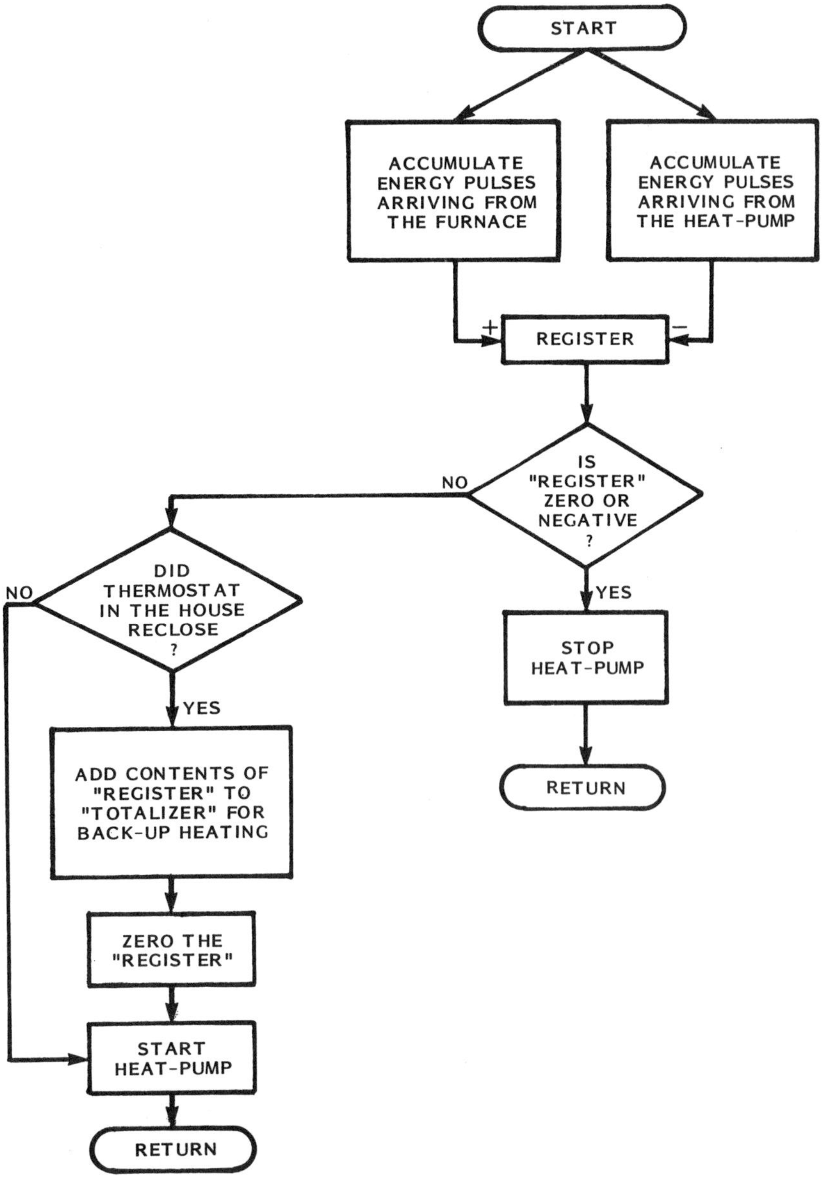

Figure 63. Flow diagram of control for heat-pump to MIMIC a furnace.

heat from indoor air. This heat deficit or loss can be measured in a manner similar to that for heat output and stored in a separate register. This deficit is to be subtracted later from the measured heat output to evaluate the net performance of the heat pump.

Detailed evaluation of performance is possible by analyzing the records on a cycle-by-cycle, periodic, or seasonal basis. A different set of temperature sensors and different integrator is usually required to keep track of the energy consumed during a defrost, because the temperature rise observed during a normal heating cycle reverses and becomes a temperature drop, which the normal integrator cannot respond to. The second integrator would be brought into play only during the defrost cycle.

5

Setting Up a Versatile Laboratory Test Station

MEASUREMENT OF PERFORMANCE BY SIMULATION

The performance of a solar system depends on properties and characteristics of the components and controls, on weather effects, and on use. Although the intrinsic properties and characteristics of a system are determined by the design selected, the effects on performance of changing many of the design parameters cannot be readily predicted by the designer. Weather is generally so variable that it is nearly impossible to repeat a series of tests under "identical" weather conditions, except possibly those in areas of the world with virtually cloudless skies for weeks at a time. Usage of systems is another variable with a nearly infinite number of variations. The investigator's challenge is to design an experiment to evaluate system performance where the performance itself is affected by a great number of variables as well as by use and time of use of a solar system.

A practical way to discipline the large number of the variables is to test a system under a realistic use pattern that is repeated over a period of time that is long enough for the effects of weather on performance to become evident. For example, a solar water-heating system is installed in a laboratory and is equipped with a controller capable of drawing a given volume of water from the system at various times of the day and repeating the same draw schedule day after day. If several test stations are made available, different systems can be operated under the same weather conditions and the same draw schedule. Alternately, several units of the same type of system may be adjusted to various control settings, or to different component sizes or characteristics, and tested simultaneously under the same weather and the same draw schedule. This test method is particularly suitable for comparison testing of complete systems and for testing against a "standard" or "minimum" performance requirement.

Weather, including sunshine, also can be simulated, but generally simulation facilities are used for a small sample or a single solar panel rather than for a complete set of collectors connected to an operating system. For a test of performance components to be useful, radiation intensity and wavelength, temperature, wind speed and direction, and flow of heat transfer fluid must be controlled. A weather simulation facility is not particularly useful for providing such control.

Simulation testing by use of a specific draw schedule has been used for many years to obtain performance information on domestic water heaters for residential and commercial use. The Canadian Standards Association (CSA) produced Standard C191-1973 "Performance Requirements for Electric Storage—Tank Water Heaters," which, in addition to water heater construction standards, contains a test procedure with a prescribed draw schedule. For each size of water heater, a specific draw schedule is given in terms of liters to be drawn at specific times. The original data were obtained from a hot-water-use survey conducted across Canada a number of years ago. As expected, there is considerable range of daily total hot water use, and the total for each household varies from day to day. The draw schedule selected for the CSA Standard was representative of most of the population but did not include all of the high users. Water heaters that are capable of meeting the test-draw schedule are therefore expected to fail to meet the needs of a certain number of high-volume users of hot water. Over the years, the hot-water-usage pattern and volume have probably changed because of the higher number of working mothers, fewer family formations, smaller families, and higher price of energy. To obtain a meaningful draw schedule, the hot-water usage (on an hourly basis, for example) of as many households as possible must be measured or the information must be collected by a questionnaire. Under the latter method, the householders keep track of their hot water use and time of use, and their data are used, along with suitable assumptions, to estimate the volume used during each hour of the day. A typical draw schedule

TABLE 7
Typical Draw Schedule

Time	8:00	9:00	10:00	11:00	11:30	12:00	13:00	14:00	15:00
Litres	41.0	41.0	45.0	91.0	82.0	50.0	13.5	13.5	13.5
(GAL)	(9)	(9)	(10)	(20)	(18)	(11)	(3)	(3)	(3)

Time	16:00	17:00	18:00	19:00	20:00	21:00	23:00
Litres	4.5	13.5	9.0	9.0	13.5	9.0	4.5
(GAL)	(1)	(3)	(2)	(2)	(3)	(2)	(1)

is shown in Table 7. The total draw in the 24-hour sample period is 454 liters (100 gal).

A water heater is considered to fail the test if the outlet temperature at any time drops below a predetermined minimum acceptable level. Normally, the test should simulate a realistic situation, rather than an extreme case of high use in which the water-heater system could not possibly satisfy the demand. In such a case, the performance failure would be uncontrollable, and it would be difficult to evaluate the amount of energy deficit in the shortfall.

CONTROL EQUIPMENT

Simulations of hot water use can be conducted in a variety of ways; some are more accurate than others. For example, if we assume that a solar water-heating system is installed at a test location, the cold water supply line is connected to the system, and power is provided to the controls and to the backup tank or heater, then the simulation consists of merely opening the outlet valve at the appointed times, discharging the prescribed volume of hot water, and noting the average temperature difference between the outlet and inlet. The prescribed volume can be obtained either by timing the draw through a pre-set outlet orifice or by observing on a meter the flow of cold water going into the system and closing the outlet valve at the appropriate reading.

The former method is less accurate than the latter, mainly due to the fluctuation in water pressure of the mains affecting the flow rate. If a water flow meter with a pulse-output capability is used, then a hard-wired programmer can be assembled from electromechanical timers, stepping relays, and counters and pre-set to the number of liters for each draw. A much less costly alternative is to use a microcomputer in which the number of litres and the starting time of each draw are stored in memory. The microcomputer opens and closes a solenoid valve, letting out the required volume of water according to the schedule. If the solarimeter output and kWh consumption informaton is available in pulse form, as discussed in Chapter 1, the microcomputer can be programmed to total the number of contact closures of the flow meter and store the readings for any required time intervals such as 1 hour, 24 hours, etc. The advantage of using a microcomputer is the flexibility that its programming provides. In fact, the same hardware can be used for this mode of operation as was used for the MIMIC mode described in Chapter 4 (see Figure 54). The only difference is that the telephone lines to the house are not needed. Instead, a draw schedule is entered into the microcomputer's memory, and the device is programmed to open the valve, keep track of the water being drawn, and close the valve after the appropriate volume is drawn. The microcomputer that was described in Chapter 4 to conduct the MIMIC operation can be used for controlling this simulation. All that is required is a new program. A detailed description and flow diagram of this program is presented in Appendix C.

AUXILIARY EQUIPMENT

When testing water heaters by simulation it is important to have a regulated water pressure, regulated power supply, and controllable and stable ambient temperature in order to obtain repeatable results. Since incoming water temperature may vary with the seasons, compensation for inlet water temperature must be made to obtain usable measurements of temperature rise across the water heater. In large buildings, the cold water supply line may pick up heat along the piping, especially when the draws take place infrequently. To prevent heated supply water from adversely affecting repeatability, a bleed to the cold water supply must be provided just ahead of the test station, as was shown in Figure 22, allowing just sufficient flow to maintain the cold water inlet acceptably close to the actual incoming water temperature.

It is desirable to monitor the cold water supply temperature and the system outlet temperature, to check that the controls on the solar system are operating properly, and to check that there is no loss of stored heat via the collector during the night. It is helpful to have a stand not too far from the water tanks to support the test collectors. These should have an adjustable angle of elevation and should be oriented due south. A solarimeter should be mounted near the top of the stand at the same angle to the sun as the collectors.

THE COMPLETE TEST STATION

A very convenient arrangement for the instrumentation is to mount all components on an instrument rack and to run the control, power, and data wiring to the various parts as required. Once made, the connections between instruments and most sensors and controls need not be disturbed when the test water heater is replaced with another. In Figure 64, the location of components of the instrument package is shown, and the connections to the various components are indicated. This complete set-up can be used for either the simulation or MIMIC modes of operation as determined by the microcomputer program in use.

The other parts of the test set-up that can be considered to be "fixed" are: the test rack and piping from the roof, the back up water heater, the cold-water supply and drain, and a suitable power supply. Temperature transducers and most thermocouples can be installed permanently on these fixtures. Changing solar water heaters then involves a minimum of disruption to the test set-up.

When more than one system is tested at the same time, it is a good idea not to draw water at the same time from all systems to prevent the water mains pressure from dropping severely during the draws.

The test station can be left operating unattended for long periods of time, according to the frequency of retrieval of records required.

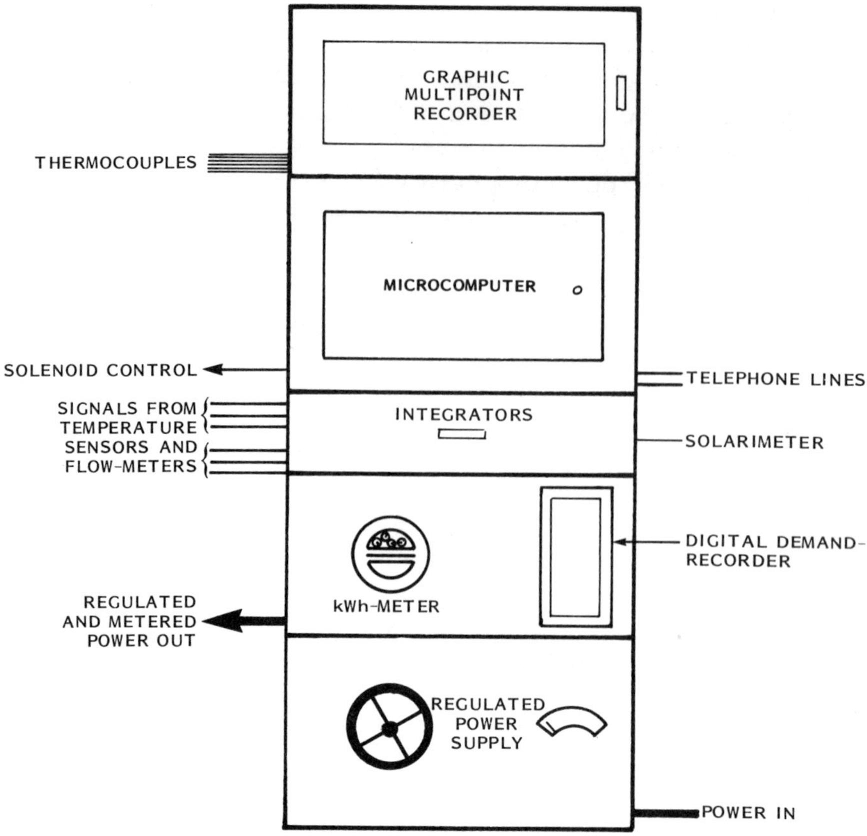

Figure 64. Instrument package for a solar test station.

Appendices

Appendix A
Heat Flow Integrator

DESCRIPTION OF CIRCUIT AND OPERATION

The heat flow integrator is capable of multiplying two variables such as flow rate and temperature rise in real time to produce a signal proportional to energy flow. The signal is in the form of pulses whose frequency varies in proportion to the flow of energy, making integration with respect to time a simple matter of counting pulses over the desired length of time. The integrator effectively pre-processes temperature and flow information, thus eliminating the need for separate recording of the temperature-difference and flow-rate information for later computer processing. The schematic of the heat flow integrator is shown in Figure A-1. All components can be mounted on a single circuit board, about $5'' \times 8''$ in size. A DC regulated power supply is required for the integrator, the temperature sensors, and the switch contacts on the flow meter. Since a variety of integrated circuit components are available to perform the signal processing, only a general description of the requirements will be given.

The temperature transducers are connected in a bridge circuit; when the bridge is unbalanced by development of a temperature difference, current flows in direct proportion to the magnitude of temperature difference. The voltage associated with this current flow is applied to a converter that changes the voltage to a series of pulses whose frequency is proportional to the amplitude of the input voltage—and, consequently, to the temperature difference. This is the data signal.

The pulses generated by the water-flow meter are input to a Schmitt trigger, which in turn controls the operation of a monostable (one-

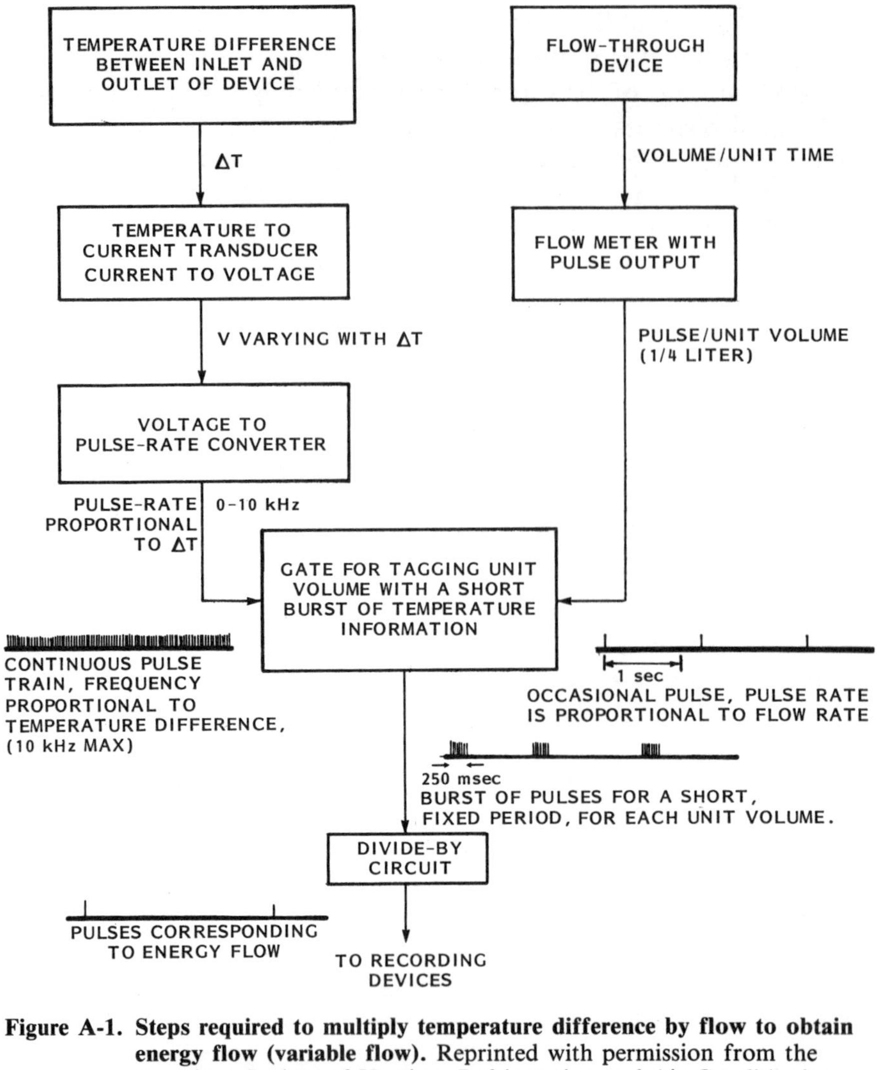

Figure A-1. Steps required to multiply temperature difference by flow to obtain energy flow (variable flow). Reprinted with permission from the American Society of Heating, Refrigerating and Air Conditioning Engineers, Atlanta, GA.

shot) multivibrator. (The function of the Schmitt trigger is to prevent spurious cycling of the multivibrator by circuit noise or contact bounce in the flow meter.) The multivibrator produces a relatively long pulse—it may be on the order of 250 msec. compared to the approximately 0.1 msec. duration of a single signal pulse. The multivibrator pulse serves a gating function; during its ON period it enables the data signal to be applied to a dual flip-flop circuit; during its quiescent period, flow of the data signal is inhibited.

The first stage of the flip-flop divides the pulses of the data signal by 2 and also shapes the signal to a form compatible with a programmable counter or "divide-by-N" circuit, where N can be selected from 3 to 9999 by setting switches. The output signal can be divided by 2 again by feeding it to the second half of the dual flip-flop. The output of this circuit can be used to drive a transistor capable of operating an output relay, which in turn operates a recording device.

Calibration Procedure for the Heat Flow Integrator (HFI)

The most convenient method of calibrating the heat flow integrator circuit is to connect a power supply and the temperature transducers to the printed circuit board and simulate the operation of a flow meter with a pulse generator. A 44-pin card-edge connector can be used to facilitate contact with the board. A terminal strip can be used to connect external circuits to the card-edge connector. Other items of equipment required are: a pair of Dewar flasks (1-liter volume), a pair of calibrated thermocouples or remote temperature detectors, a readout device capable of 0.1°C or better resolution in measuring temperature difference (absolute temperature is not required), an electronic frequency counter, a pulse generator, and, if possible, a DC-coupled oscilloscope.

The sensors must be identified "cold" and "hot" and must not be interchanged. The calibration applies only to positive temperature differences ("hot"–"cold" temperature).

Temperature to Frequency

The first test temperature difference is zero, which is obtained by immersing both hot and cold sensors in oil or water (depending on the type of sensor being calibrated) in one of the Dewar flasks. Allow sensors to stabilize in the bath for 5 minutes. With power applied to the circuit board, adjust the circuit components until the output frequency of the circuit itself just approaches zero. A frequency of up to two Hz is acceptable.

Next, provide about 10 temperature differences 10°C to 15°C apart (or 1 to 1.5°C apart for the lower-range version) over the full operating range by adding hot or cold oil or water (ice, salt). After the readings stabilize at each temperature condition, note the temperature difference and the circuit frequency. By plotting frequency versus temperature or by using linear regression, obtain the slope and intercept expressed as A and B, respectively, in the equation following:

$$F = A(\Delta T) + B \tag{A-1}$$

where: F is the circuit frequency (Hz)
ΔT is the temperature difference (°C)
A is a circuit board constant, typically around 100 Hz/°C
B is a constant (Hz) which we want to be near zero

If the data points do not line up along a very straight line, probably the circuit did not stabilize, or an error was made in taking the readings. The correlation coefficient, which gives a measure of the quality of the linear fit, should be greater than 0.98. If smaller, repeat this step of the calibration.

Since, in our application of this circuit, we require that no pulses be produced at zero temperature difference, we must force B to be zero. This can be done by re-adjusting circuit components to create an offset of the input bridge at any temperature difference (except zero) so that:

$$F_{new} = F_{old} - B \tag{A-2}$$

where: F_{new} is the frequency after adjustment
F_{old} is the frequency before adjustment
B is the intercept, in Hz, equal to the offset in frequency that is required

Recheck the calibration at different temperatures and re-plot or recalculate the regression line. Once B disappears, the relationship becomes:

$$F = A(\Delta T) \tag{A-3}$$

$$A = \frac{F}{\Delta T} \tag{A-4}$$

where frequency is linearly proportional to temperature difference with no offset from the origin.

Adjustment of the ON Time of the Gate

A pulse generator may be used to substitute for the contacts of the flow meter. A unipolar positive pulse of sufficient amplitude can be used to trigger the Schmitt trigger. The Schmitt trigger's output pulse then operates the monostable multivibrator, whose output width is controlled by the values of adjustable circuit components. The ON time must be shorter than the period between the flow pulses. For example, if flow pulses are produced at a maximum rate of 0.5 Hz, then the period between pulses would be 2 seconds, and the ON time could be chosen to be 1.8 seconds or less.

Setting the "Divide-By" Circuit

The divider is programmed by DIP switches. The range of division is from 3 to 9999 and is generally selected to meet the requirements of the recording device and of the driver relay on the circuit board. For example, the maximum counting speed of a digital demand recorder is typically 50 counts per minute. The maximum counting speed of the circuit board driver relay is about 600 counts per minute; it would certainly not cause any limitations while driving the digital demand recorder. The "divide-by" factor must be selected so that the maximum input rate of the recording device is not exceeded. The energy associated with each pulse produced by the heat flow integrator is:

$$E = V \frac{\Delta T_{max}}{F_{max} \times P} c \times N \times k \qquad \text{(A-5)}$$

where V = volume associated with a flow-pulse
P = ON time of the gate
c = specific heat of the fluid
N = divider switch setting
k = is a constant—2 for edge-triggered totalizers and 4 for pulse-triggered totalizers, necessary to account for the division by 4 caused by the dual flip-flop.

Substituting $\frac{\Delta T_{max}}{F_{max}}$ in (A-5) by $\frac{1}{A}$ (from equation A-4) then:

$$E = \frac{V}{AP} cNk \qquad \text{(A-6)}$$

Since N is selectable to any desired value that satisfies the counting speed of the drive relay and recorder and the capacity of any totalizing counter, it can be selected to be a convenient portion of a kWh, such as 1/10 or 1.

Example: Calculate the minimum value of N to satisfy the limiting conditions.

Suppose the demand recorder used is capable of recording, at the most, 50 pulses per minute, or 1 pulse per 1.2 seconds. Assume that the calibration factor of the circuit (temperature-to-pulse rate conversion is 100 Hz/°C. Assume also that the ON time of the gate is ½ second and the flow meter produces 1 pulse per liter of water flowing. Calculate the value of N which will produce a satisfactory pulse rate. For a domestic water heater typical parameters might be:

$V = 1$ Liter
$A = 100$ Hz/°C
$P = 0.5$ sec.
$c = 0.00116$ kWh/liter°C
$\Delta T_{max} = 50°C$
Maximum flow-rate = 16 liters/minute (960 liters/hr)
Maximum recording speed = 1 pulse/1.2 seconds

The limiting condition turns out to be the maximum recording speed of the recorder. Each time a flow pulse initiates a ½-second operation of the gate, it must be ensured that *at most* 1 energy pulse is produced. If two or more pulses were produced during the half second period, the recorder would not be capable of discerning them. In this case, the *minimum* factor Nk to divide by is equal to the total number of pulses produced during the gate ON time at the highest temperature difference that will be encountered. The number of pulses admitted by a single operation of the gate will be:

$$AP\ \Delta T_{max} = 100 \times 0.5 \times 50$$
$$= 2500 \text{ pulses}$$

In order to have not more than 1 output pulse produced, a dividing factor (Nk) of 2500 or higher must be used. If a factor of 2500 is selected, since $k = 4$, then N must be 625, and the amount of energy represented by one output pulse at the recorder will be, as in equation A-6:

$$\text{Energy/pulse} = \frac{V}{AP}\, cNk$$

$$= \frac{1}{100 \times 0.5} \times 0.00116 \times 625 \times 4$$

$$= 0.05814 \ \frac{\text{kWh}}{\text{pulse}}$$

Since the dividing factor Nk can be chosen higher than the minimum value of 2500 calculated above, let us determine the value of Nk required for each pulse recorded to represent 0.1 kWh. By substituting energy pulse with 0.1 in equation A-6, and rearranging:

$$\text{Nk} = \frac{\text{Energy/Pulse} \times \text{AP}}{\text{c V}} = \frac{0.1 \times 100 \times 0.5}{0.00116 \times 1} = 4310$$

Since k=4, then N required is 1078.

If a totalizing counter is to be used and read weekly, we can calculate the minimum number of digits that the counter requires so as not to overflow during the recording period. By assuming that the system will operate continuously at the maximum flow and temperature difference, we can calculate the maximum energy delivery capability thus:

$$\begin{aligned} \text{Maximum Output} &= \text{Max flowrate} \times \text{c} \times \Delta T_{max} \\ &= 360 \times 0.00116 \times 50 \\ &= 20.9 \text{ kw} \end{aligned}$$

Each hour, the system would deliver at most 20.9 kWh, and the instrumentation would produce ten pulses per kWh, or 209 per hour. At this rate, after one week, the following number of pulses would be produced?

$$\begin{aligned} &209 \times \text{number of hours per week} \\ &= 209 \times 7 \times 24 \\ &= 35{,}112 \text{ counts} \end{aligned}$$

A 5-digit totalizing counter would be adequate for weekly readings.

HEAT FLOW INTEGRATOR FOR SPECIAL CASE
Flow rate is constant value or zero

An "indirect method" of measuring heat output can be applied to air- or water-based heating systems that have forced circulation and a rate of mass flow essentially constant over time or zero. The indirect method involves (1)"calibration" with the aid of an electric heating element in the air stream or water stream to establish the temperature rise caused by a known heat input and (2) relating the temperature rise produced by the normal heating plant to a corresponding rate of energy output, assuming a linear relationship between power input and temperature rise. Recording of energy output consists of converting temperature rise to the linear pulse rate and multiplying by zero or one, depending on whether the blower or pump is OFF or ON. See schematic diagram of the circuit shown in Figure A-2.

A simplified version of this circuit can be used to perform the integration. The temperature sensing and conversion is identical, but the flow signal (a DC voltage indicating pump or blower-motor status) can be used directly to gate the dual flip-flop. The "divide-by" circuitry is still required, and an output relay is used in the same manner as previously.

Calibration Procedure for the Temperature Sensors

The temperature sensors and power supply are connected to the circuit board, as previously described in this appendix. The temperature-to-frequency calibration procedure is carried out exactly as before, so that frequency will be linearly proportional to temperature difference without any offset.

Since the gating operation consists of lengthy ON periods (several minutes) and OFF periods of longer or shorter duration, the output data will appear only during blower or pump ON periods. The calculation of a suitable N factor by which the number of pulses must be "divided" is very similar to the previous calculation in this appendix, for example:

Suppose that during normal operation, a forced-air oil-fired furnace produces maximum temperature rise T_{max} of 60°C. If the temperature-to-pulse-rate conversion factor A is 100 Hz/°C as before and the pulses are to be recorded during the entire blower-ON period, then the maximum frequency generated by the integrator will be:

$$A\ \Delta T_{max} = 100 \times 60 = 6000\ \text{Hz}$$

If a recorder can accept 1 pulse every 1.2 seconds, then the above frequency must be divided by a factor that gives a frequency of 1/1.2

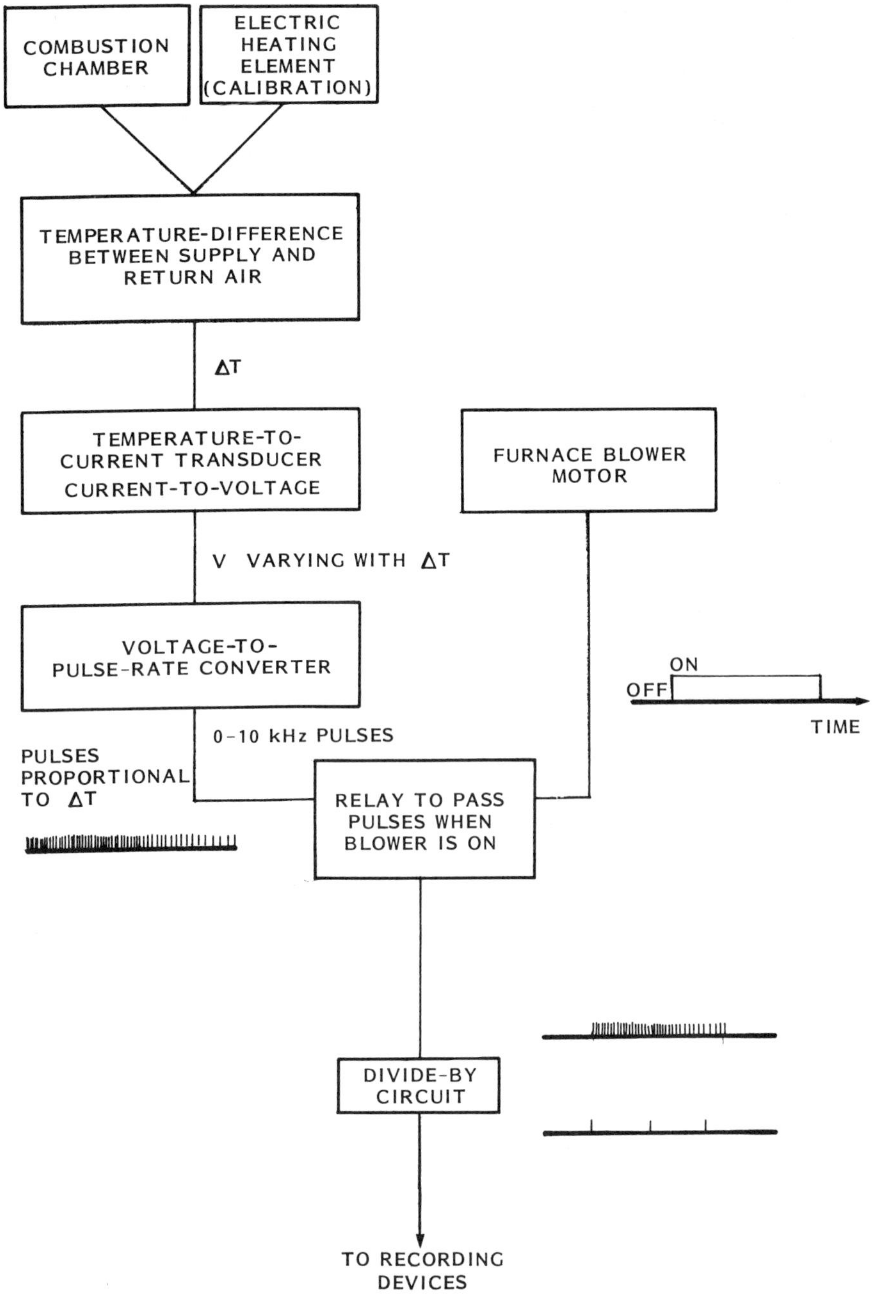

Figure A-2. Steps required to multiply temperature difference by flow to obtain energy flow (constant flow).

or 0.833 Hz or smaller. The factor kN must be 7200 or larger, and N must, therefore, be 1800 or larger.

If the value N=1800 is selected, a calibration will yield the amount of energy that each recorded pulse represents. For example, suppose that a 20 kW element produces a temperature rise of 30°C. The output frequency of the temperature-rise-to-pulse-rate converter will be:

$$A\ \Delta T = 100 \times 30 = 3000\ \text{Hz}$$

Downstream of the "divide-by kN" circuit the frequency will be 3000/7200=0.4167 Hz.

During 1 hour, 20 kWh will be delivered, and the following number of pulses will be recorded:

$$\text{Frequency (Hz)} \times \text{Period (seconds)} = 0.4167 \times 3600 = 150\ \text{pulses}$$

Therefore: each pulse represents 20/150=0.133 kWh.

Higher values of kN may be selected to increase the amount of energy represented by each pulse to any convenient value such as ¼ kWh or 1 kWh.

Appendix B

Microcomputer Control of MIMIC Operation

A microcomputer can be used as a very economical means of controlling a test water heater so as to MIMIC another in a remote location. With additional programming and spare storage capacity, the microcomputer can be used to store readings of flow, energy used, and sunshine available for manual reading, or for automatic recording, printing, or processing.

Since a large number of single-board computers are now available, the description of the circuit configuration and the program will be given in block diagram form for general application. The microcomputer must have a 1K ROM monitor capacity, 1K of RAM storage with battery back-up, and an interface board with 7 digital inputs and one output relay. A key pad for input and an 8-digit display for memory address and data are adequate. The configuration is shown in Figure B-1. The computer program is structured in two parts: the main program for control and an "interrupt subroutine" for keeping time and for taking readings. Details of the main program and subroutine are discussed separately below.

The MIMIC program, whose flow diagram is presented in Figure B-2, monitors digital input channel number 6, which receives the contact closures from the remote location corresponding to specific increments of hot water drawn from the conventional electric water heater in the house. The computer's internal clock counts every 15 minutes, saves the total pulse count of channel 6 over that period, and updates the draw schedule for the next 15-minute period with the data received. At the beginning of the next time period, the data received becomes the amount of water scheduled, which is compared with the number

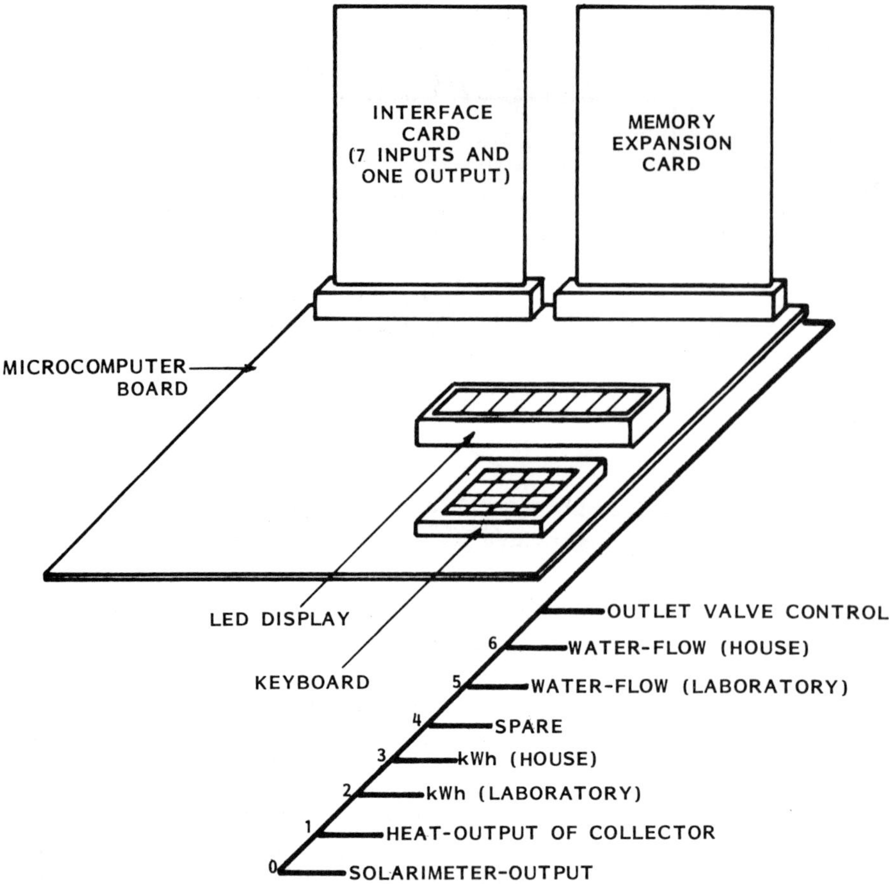

Figure B-1. Minicomputer board and interface to control the MIMIC operation.

of flow pulses received on channel 5 from the flow meter on the test system in the laboratory. If the total number received from the house is greater, the outlet valve is opened under computer control via the control relay. As soon as the number of pulses received from the laboratory system match, the computer closes the outlet valve and stands by until the total number of counts for the next 15 minutes is established. The computer keeps track of overall time and at midnight takes the daily totals of the variables connected to channels 7, 5, 4, 1, and 0. Data recorded may be: energy delivered by the solar collector, kWh used at the house, kWh used at the laboratory, hot water drawn

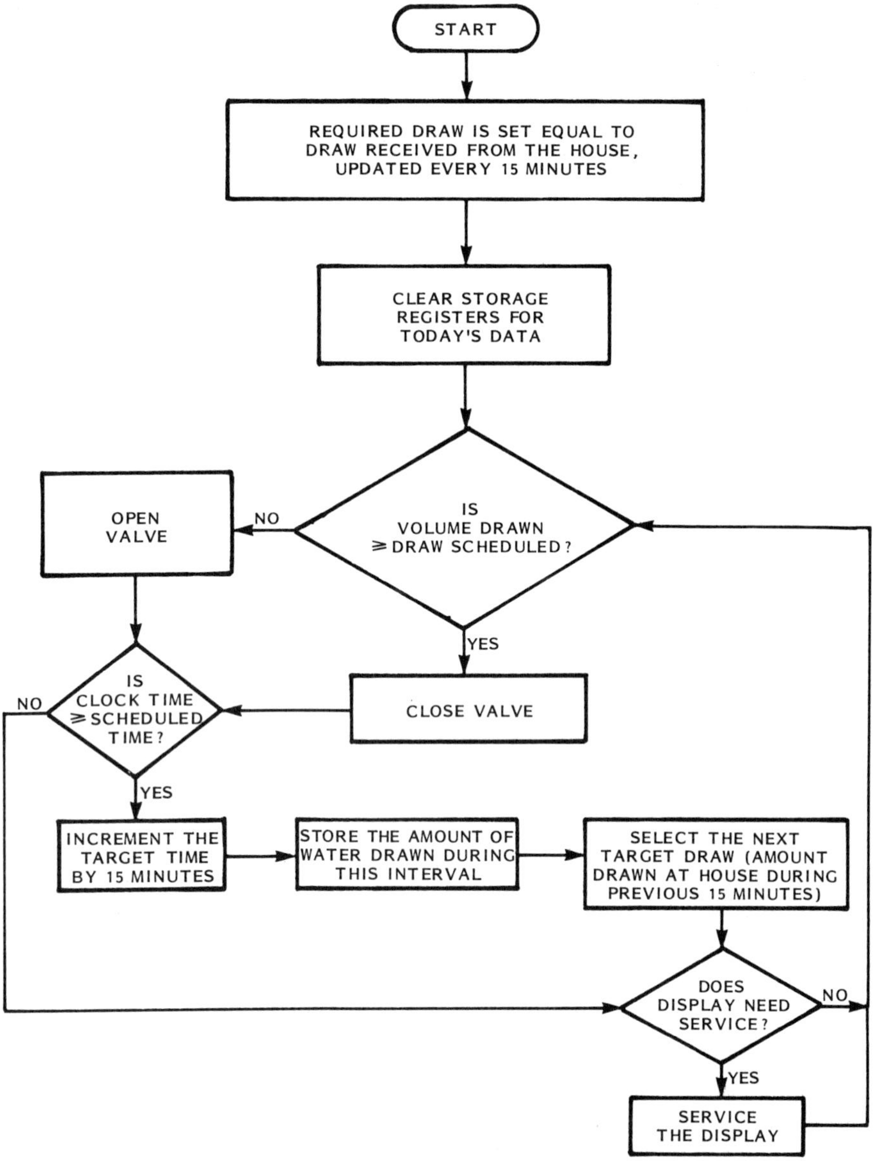

Figure B-2. Flow diagram of MIMIC control program.

at the laboratory, and amount of incident solar energy available. Channels 2 and 3 are used to record 15-minute totals for the selected data over one complete day. The flow diagram of the recording subroutine is shown in Figure B-3.

Although the subroutine can be executed every 16.6 msec. (once per 60-Hz cycle), a scan of the input lines for any change in every other cycle is adequate. If a change from the previous scan has occurred, the appropriate registers are incremented. Every 3600 cycles, the clock time is advanced 1 minute. Every 15 minutes, the counts accumulated on digital inputs 2 and 3 over the previous 15 minutes are stored.

At the end of each day, the daily totals for the previous 5 days and the total for the day ending are reassigned a storage location and the oldest data (from the 6th day) is lost. The 15-minute values stored from channels 2 and 3 are assigned the same storage locations as the previous day's data, overwriting the old data with new.

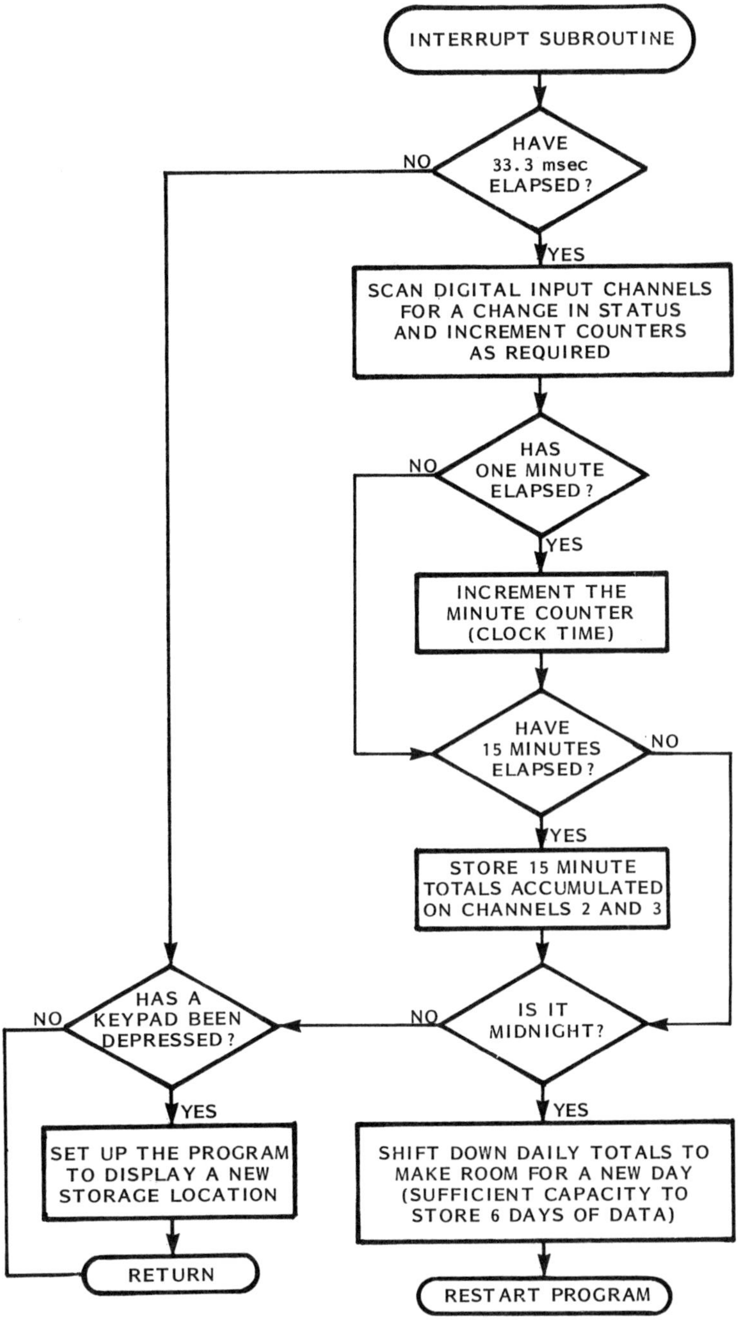

Figure B-3. Flow diagram of recording subroutine.

Appendix C

Microcomputer Control of Simulation Operation

The test method for subjecting water heaters to a given draw schedule can be executed by programming the microcomputer described in Appendix B with the routines described below. This example illustrates the flexibility and convenience of a programmable microcomputer for this application.

The main program stores the draw schedule, which can be changed manually to suit. The "interrupt subroutine" for storing daily totals and 15-minute data on two channels is identical to the one described in Appendix B for the MIMIC operation.

Refer to the flow diagram in Figure C-1 for the program to control the "simulation" operation. This schedule of draws is entered into storage. A set of 16 specific times and 16 specific volumes is typical of what may be used to simulate a test use pattern of hot water. The computer starts at midnight by clearing the storage registers for the current day's total counts. The pointer is set at the first value of the draw schedule, indicating the amount of water to be drawn and the time at which the draw is to take place. At the specified time, the computer sends an "open" signal to the solenoid valve and checks that the volume of water drawn is the same as or greater than the amount specified in the schedule. When this occurs, the computer sends a "close" signal to the solenoid. As a double-check on the accuracy of the control, the computer also stores the amount of water drawn during each draw interval, so that it can be checked against the required draw schedule after the test is completed. This check on the control system is a good idea for security and valuable for ensuring the integrity of the data, because discrepancies may occur due to malfunctions of the solenoid valve, possible programming error, or power interruptions.

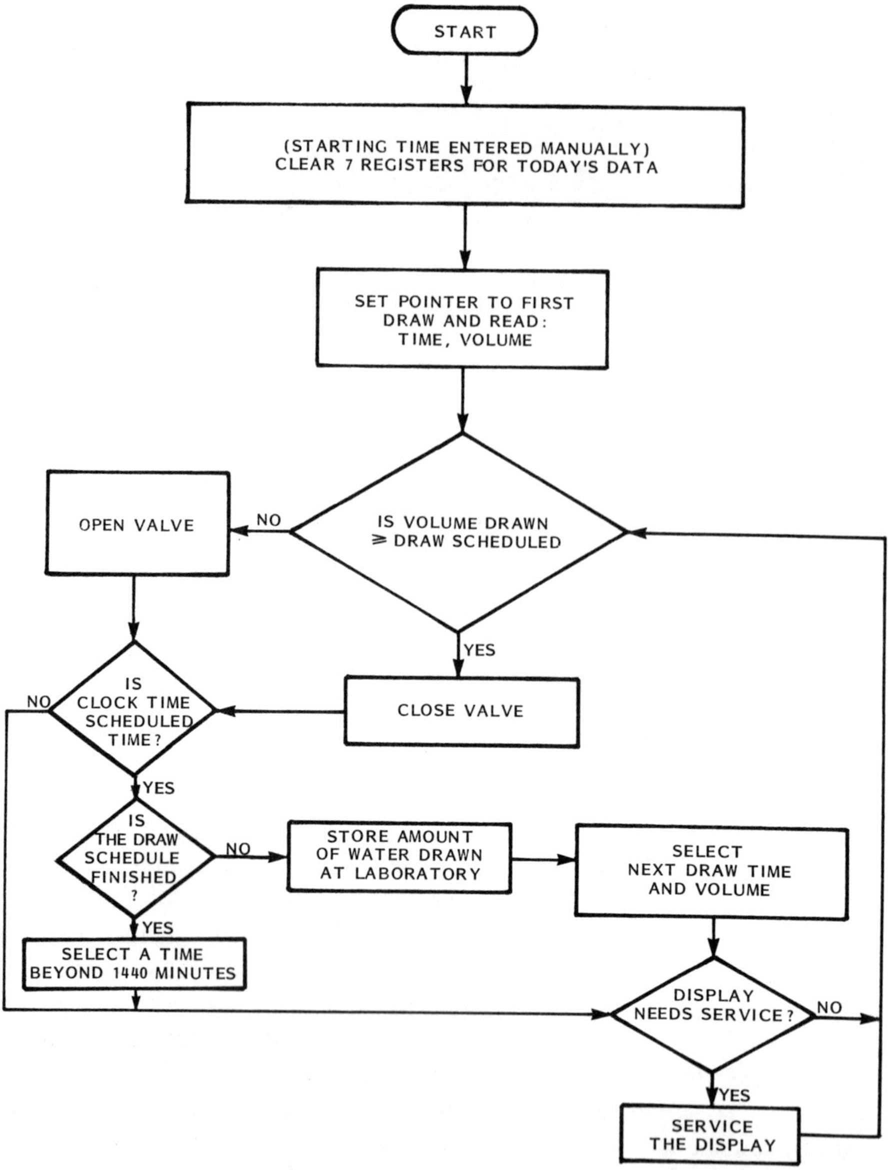

Figure C-1. Flow diagram of simulation control program.

References

INTRODUCTION

1. *ASHRAE Handbook, 1981 Fundamentals,* p. 1.2.
2. Stricker, S., "Energy Inputs and Heat Losses in Residences," *Ontario Hydro Research Quarterly, Vol. 28, No. 4,* pp. 19–23, 1976.
3. *ASHRAE Handbook, 1981 Fundamentals,* 1981, p. 8.12.

CHAPTER 1

4. *ASHRAE Handbok, 1981 Fundamentals,* Chapter 27.
5. Abbott, C. G., "Solar Variation and Weather," *Smithsonian Misc. Collection,* Vol. 146, No. 3, Oct. 18, 1963, pp. 2–9.
6. *ASHRAE Handbook, 1981 Fundamentals,* 1981, p. 15.3
7. *ASHRAE Handbook, 1981 Fundamentals,* 1981, p. 15.5
8. *ASHRAE Handbook, 1981 Fundamentals,* 1981, p. 15.8

CHAPTER 2

9. *ASHRAE Handbook, 1981 Fundamentals,* p. 13.15
10. *ASHRAE Handbook, 1981 Fundamentals,* p. 13.18

CHAPTER 3

11. Stricker, S., "Solar System Performance Evaluation Techniques—The MIMIC Principle," *ASHRAE Journal,* Nov. 1979, p. 67

12. *ASHRAE Handbook, 1980 Systems,* p. 20.4

CHAPTER 4

13. Stricker, S., "Solar System Performance Evaluation Techniques—The MIMIC Principle." *ASHRAE Journal,* Nov. 1979, p. 65.

Index

P

R

S

T